URBAN TRANSPORT
WITHOUT THE HOT AIR

URBAN TRANSPORT
WITHOUT THE HOT AIR

STEVE MELIA

Published by
UIT Cambridge
www.uit.co.uk

PO Box 145, Cambridge CB4 1GQ, England
+44 (0)1223 302 041

Copyright © 2015 UIT Cambridge Ltd. All rights reserved.
Subject to statutory exception and to the provisions of relevant collective licensing agreements, no part of this book may be reproduced in any manner without the prior written permission of the publisher.

First published in 2015, in England

Steven Melia has asserted his moral rights under the
Copyright, Designs and Patents Act 1988.

Cover design by Jonathan Pelham
Interior design by Jayne Jones

ISBN: 978 1 906860 27 1 (hardback)
ISBN: 978 1 906860 26 4 (paperback)
ISBN: 978 1 906860 47 9 (ePub)
ISBN: 978 1 906860 67 7 (pdf)
Also available for Kindle.

Disclaimer: the advice herein is believed to be correct at the time of printing, but the authors and publisher accept no liability for actions inspired by this book.

10 9 8 7 6 5 4 3 2 1

Contents

    Preface vi
    Foreword by Prof Phil Goodwin vii
1   The myths of urban transport... 9

PART I: Myths and problems.. 12
2   The problem 13
3   "There as been a war on the motorist"................................. 19
4   "Roads and airports benefit the economy" 29
5   "All we need is better public transport"............................... 38
6   "Car ownership isn't a problem – only car use" 49
7   "You'll never get people over here cycling like the dutch"....... 60
8   "The car can be a guest in our streets"................................. 79
9   "We are building too many flats"... 92
10 Summary: myths, values and challenges 110

PART II: Sustainable solutions 111
11 Four options for traffic in towns.. 112
12 European cities: inspiration and similarities 123
13 Carfree developments... 149
14 London: the politics of bucking the trend......................... 163
15 Progress in other British cities... 184
16 What sort of cities do we want? 211
17 What can I do? 231
    Endnotes 240
    Index 257

Acknowledgement

I would like to thank everyone interviewed for this book, named and 'off the record'
– **you know who you are.**

Preface

This book is for anyone with an interest in transport: if you work in it, study it, or just use it and have ever stopped to ask "Why?", this book is for you. It mainly concerns the movement of people in urban areas, where three-quarters of us live, but it will also make occasional reference to other issues, like aviation and rural transport.

Part I focuses on myths and problems, leaving discussion of potential solutions to Part II. Both parts are written from a UK perspective, drawing on international evidence, though we have tried in editing this book to explain or avoid terms which would only be understood in Britain. Some of the analysis refers only to England, or to England and Wales, because comparable data is easier to obtain for those countries. The discussion of housing and population growth focuses particularly on southern England, where the problems are most acute, but some comparisons are drawn with other parts of the UK, where circumstances and policies have been rather different.

Some evidence comes from my own research, some by colleagues at the University of the West of England (UWE) and by many others. I will generally avoid naming authors in the main text, because there are so many and because I want to avoid the off-putting conventions of academic writing (Turgid et al 2015). All the sources are listed in the notes at the end of the book. If you are wondering about my interpretation of something, take a look at the original evidence and form your own view. Most of the sources are available online, although some of the academic ones require a subscription.

Foreword

Steve Melia is the man who first used the phrase 'filtered permeability' – a bit of a mouthful, but an extraordinarily useful planning tool of how to enable swifter and more convenient through routes to cyclists and pedestrians than cars in residential neighbourhoods. He is a lot closer to transport practice than most academics, and a lot closer to research than most practitioners, and entirely dedicated to the proposition that if you have things to say, you ought to engage in debate and argument. So he has written a challenging book. In identifying and confronting a number of deeply entrenched myths, he will no doubt make a few people cross, and more a bit worried – 'I wouldn't have quite put it that way' is a response that most readers, I guess, will find at least in a few places in his argument.

Steve takes an eclectic approach to evidence. "This book does not pretend to be value-free", he writes. There is no implied hierarchy of official over unofficial, national government over local government, academic over non-academic, or quantitative over qualitative. Evidence is what you can see when you are looking, which is relevant to the big problems he confronts. One of the really interesting things about the debate on transport policy is how firmly established is the theme of 'evidence', and how limited is the common ground of what constitutes evidence, and where to look for it.

As always, the treatment does raise some unanswered questions. For me, the biggest one – maybe the subject of a future book? – is a problematic issue of the relationship between different elements of the sustainable transport agenda that Steve advocates. To emphasise, as he does repeatedly, that a policy package as a whole needs to be coherent and internally consistent, is entirely in tune with the needs of the time, and a necessary counter to piecemeal token policies. But that leaves unanswered the question of the sequence of implementation, since with the best will in the world, no local or national government institution can do everything at once. So what is first, and what is the best order for the rest?

Phil Goodwin
Emeritus Professor of Transport Policy,
University College London and University of the West of England

CHAPTER 1
The myths of urban transport

Transport is part of the identity of every country, every town and city. How people travel tells you much about who they are, just as their buildings, their history and their politics do. And like all of these things, transport generates stories, which people tell to explain why things are as they are, or why they are not as people would like them to be. Some stories are told among transport specialists and others are told more widely, in the media, in politics and by ordinary people. Some stories contain important elements of the truth and others are more like urban myths, spread because they suit a prejudice, a legitimate desire or a vested interest. Several years of listening to these myths persuaded me to write this book.

To work out whether any of these myths have influenced you, here are a few questions drawn from the chapters in the first part of this book. If you don't already know the answers, take a guess:

1. **What happened** to the UK duty on petrol between 2000 and 2012?
2. **What happened** to the total UK tax on petrol (including VAT) over the same period?
3. **What percentage** of households in Britain have a car?
4. **What percentage** of international flights from UK airports are for business purposes?
5. **What percentage** of Manchester's commuters travel by tram?
6. **Do Germans** own more or fewer cars than Britons?
7. **Do Germans** drive more or less than Britons?
8. **What percentage** of journeys are for commuting in Britain?
9. **What percentage** of households are families with a couple and one or more children?
10. **What percentage** of dwellings in England are flats?

(Answers on the next page)

Answers	Chapter
1. Petrol duty fell by 16 per cent after allowing for inflation between 2000 and 2012	3
2. The total tax on petrol fell by 5 per cent between 2000 and 2012, allowing for inflation	3
3. 75 per cent of households in Britain have one or more cars.	3
4. 19 per cent of international flights from UK airports are for business purposes	4
5. 1.4 per cent of Manchester's commuters travel by tram[1]	5
6. Germans own more cars than Britons	6
7. They also drive more than Britons	6
8. 16 per cent of trips are for commuting	9
9. 19 per cent of households in England and Wales are families with a couple and one or more children	9
10. 20 per cent of dwellings in England are flats	9

If any of the answers surprised you, if you overestimated any of the numbers or thought Germans drove less than Britons, you have probably heard or read something influenced by these myths, and this book is especially for you. And if you are wondering why a book on transport is asking about families and flats, read on.

Chapters 3 to 9 in Part I begin with a list of myths commonly heard in the transport world, followed by 'observations', which cast doubt on some of them. I chose 'observations' rather than 'the facts' or 'the truth' because the truth is rarely straightforward and 'facts' may be interpreted in different ways. Most of the myths, like most issues in transport, involve numbers. People in the transport world like numbers: they suggest a degree of precision – hard facts instead of woolly opinions. Some of these people, including many politicians and transport economists, go further, falling for the 'calculation fallacy'.

The calculation fallacy
The belief that a calculation can demonstrate what ought to be done, with no need for value judgments.

A calculation may help to estimate the consequences of an action but the level of desirability we attach to those consequences will always depend on value judgments. Some relatively simple value judgments on technical questions may seem fairly obvious. For example, if an engineering study shows that a bridge needs reinforcement to avoid a significant risk of failure, reasonable people would all agree that it must be reinforced if it is to remain open. But how much money should be spent on strengthening bridges against the risk of freak floods? Should the bridge or the road leading to it have been built in the first place? These questions can be informed but never answered by a calculation. People – including the economists at the Department for Transport (DfT) – who pretend that a calculation can show the *right* or the *wrong* decision are often trying to disguise the value judgments behind their calculations – values which might shock many people if they knew how the system worked. Chapter 4 will look at some specific examples of this.

This book does not pretend to be value free. My starting point is a belief that maintaining the conditions for life on Earth is more important than increasing consumption or maximizing individual freedom. At a more immediate level, I have observed how life in many European cities has improved as motor traffic has been removed from streets and neighbourhoods. If you support those aspirations, you will find useful evidence in this book, but whatever your own starting point, the analysis should interest and may surprise you.

PART I
Myths and problems

CHAPTER 2
The problem

Transport and climate change

The phrase 'conditions for life on Earth', in the previous chapter, may sound over-dramatic, but those conditions are facing growing threats: from loss of biodiversity to resource depletion and particularly climate change.[2] With the possible exception of celebrity private lives, nothing has generated so much rubbish in the media and online as the science of climate change. Some of this rubbish has unfortunately found its way into *Local Transport Today*, the otherwise indispensable magazine for anyone who works in transport. For a readable summary, written by scientists who know what they are talking about (unlike most of the journalists, columnists and website authors you might have read), I particularly recommend *Climate change: A summary of the science*, published by the Royal Society and available on its website.[3]

Transport accounts for just over a quarter of the greenhouse gas emissions in the UK, second only to power generation. Figure 2.1 shows the percentages of emissions for the different modes of transport. The official figures in Figure 2.1 understate the contribution of aviation for two reasons. First, the greenhouse gases in the upper atmosphere have a greater effect than those emitted on the ground. Second, the figures are based on the fuel supplied at UK airports; however, nearly two thirds of passengers leaving the UK are UK residents, so our 'fair share' of that fuel would be nearly two-thirds, *plus* a similar amount of fuel from foreign airports. The upper-atmosphere effect is difficult to calculate – the effect on climate change is not fully understood – but is believed to be 1.3-2.0 times as powerful as similar emissions at ground level.[4] An upper atmosphere adjustment of 1.5, plus an adjustment to reflect a fairer UK resident share of fuel, would bring emissions from aviation to roughly the same level as cars.[5]

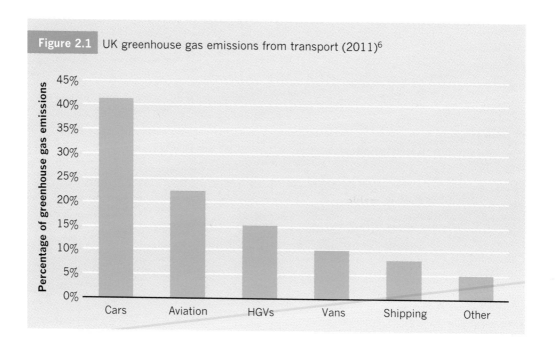

Figure 2.1 UK greenhouse gas emissions from transport (2011)[6]

Can electric cars solve the problem?

The UK Climate Change Act of 2008 sets legally binding targets for the country to reduce its greenhouse gases by 80 per cent between 1990 and 2050. This act established the Committee on Climate Change to report to parliament and make recommendations to the government about how to meet the targets under the act. In their reports on surface transport, the Committee placed great faith in electric cars. In 2009, they recommended or forecast (the distinction is blurred by the reports' use of 'scenarios') a market share for electric cars of 60 per cent by 2030, coupled with a 90 per cent decarbonization of electricity generation.[7]

A few years later, that scenario was starting to look rather optimistic. By 2013, fully electric cars represented just 0.1 per cent of new car sales.[8] And despite reasonable progress on renewables, 65 per cent of electricity was still being generated from fossil fuels in 2013,[9] while the government approved a new generation of gas-fired power stations.[10]

Electric cars are more energy efficient than petrol or diesel cars. Even where the electricity is generated by fossil fuels, electric cars generate lower emissions than the current generation of conventional cars.[11] They also reduce noise and local air pollution in urban areas. They are undoubtedly part of the solution to our transport problems – but only part.

A transition to electric cars as rapid and absolute as in the Committee's scenario raises many practical problems. Several of these relate to battery charging. Battery charging is developing rapidly, but the limited range of most electric cars remains a barrier to their uptake. As with any new technology, several competing incompatible types of charging plug still exist. The 32-amp charging points being installed by the government take 3-4 hours to fully charge the battery of a typical electric car.[12] In many places standard 13 amp domestic sockets are likely to remain the only local option for some time to come and a full charge will take between six and eight hours – fine if you are able to leave the car overnight, but not much use if you need it in a hurry. The few 'Rapid Points' which currently exist can charge the same battery in less than an hour – a considerable improvement but hardly competitive with petrol and diesel cars, which spend a few minutes in a filling station every few hundred miles.

In planning how renewable energy will supply electric cars, several analyses, including the Climate Change Committee's, have suggested that much charging would be done at home, making use of surplus capacity at night. This would be relatively easy for homes with garages or off-street parking (providing you don't get called out in an emergency overnight). But imagine if all the cars in dense urban streets like the one pictured in Figure 2.2 were charged overnight.

Figure 2.2 Dense urban streets – how would electric cars be charged?

If cables cannot be run out of each house or flat and across the pavement, we will need a roadside charging point for every car. Other technologies under development, such as under-road wireless charging pads, could solve the problem in the longer term, but the scale of the national network needed to make this viable would require far greater resources than governments are planning and it is difficult to foresee the private sector investing on this scale as long as petrol and diesel remain permitted and viable alternatives.

Hydrogen is sometimes touted as a better option to electric cars. It offers a different way of storing energy – an alternative to the battery of an electric car but one which is much less efficient. It is very bulky and leaks out of fuel tanks, making it difficult to leave a hydrogen vehicle for any length of time. Technological advances may also address those problems in future, but experience suggests that once the market has made its choice – for either electric or an alternative – the economy will then become 'path dependent': all development and marketing will be concentrated on a narrow range of options, as occurred with petrol and diesel in the 20th century.

Other environmental problems related to cars

Carbon emissions from fuel are only one of several environmental problems caused by cars. Roughly 15 per cent of the carbon emissions over the life cycle of a car are related to its manufacture and disposal,[13] and this proportion is likely to grow as the cars themselves become more energy efficient. Vehicle manufacture uses nearly 50 per cent of the world's annual output of rubber, and around 25 per cent of its glass and 15 per cent of its steel.[14] It seems that the world has sufficient reserves of lithium to supply batteries for electric cars for the foreseeable future[15] but there could be problems associated with sudden increases in demand. Other types of motor vehicle also use resources that are either scarce or limited, but the average private car makes particularly inefficient use of them, as it spends 97 per cent of its time parked.[16]

The presence of so many cars presents a particular problem for urban areas, particularly if the population density of towns and cities may have to increase in the years to come, as argued in Chapter 9. Several studies have shown how traffic severs communities, restrict the independence of children[17], contributes to obesity[18], and even affects the number of friends people have on the street where they live.[19] If no action is taken to restrain rising traffic in urban areas, it can set off a vicious circle, where conditions for walking, cycling and public transport deteriorate, pushing more people into cars and further worsening the traffic and pollution. The good news is that this can be reversed. A virtuous circle of improvement has occurred in some of the cities described in later chapters.

Population growth in an overcrowded island

A 2011 survey produced the not very surprising result that a majority believes Britain is overcrowded.[20] Three-quarters were fairly concerned or very concerned about population projections of 70 million by 2027, which will not be the high point.

The Office for National Statistics, which is independent of ministers, has a range of population projections based on different assumptions about immigration, life expectancy and birth rates. The lowest of its 10 assumptions shows population rising by over 5 million and peaking in 2043. Its central projection – the one it expects to happen – shows no end to the growth in population, reaching 86 million by 2087.[21]

As the Migration Observatory points out, many areas of immigration are outside the control of governments. The free movement of labour is a condition for access to the single European market for countries like Norway who belong to the European Economic Area, so even if Britain withdrew from the European Union, some free movement within Europe would almost certainly remain.

A combination of population growth and shrinking household sizes (until recently) has caused a growing shortage of housing in Britain. Household growth was already outstripping house building before the recession of 2008-12,[22] which made the problem much worse. The number of people on local authority housing waiting lists grew from just over a million in 2000 to nearly two million by 2012.[23] Household sizes have stabilized recently as young people stay at home longer and more adults who are not related now share accommodation. Attitude surveys suggest most of these people would prefer a home of their own – preferably one they can buy, although many of them do not believe that to be likely.[24] Whatever policies governments pursue, some greenfield building will be unavoidable if the housing shortage is to be addressed. A recent survey conducted in south-east England showed most people support home building in principle, but not if it means building on green fields, or increases traffic and congestion.[25] As conventional home building always involves one or both of those outcomes, planners and developers tend to dismiss such views as nimbyism to be overcome or ignored. Ministers have tried to argue that "all we need to do is build on another 2-3 per cent of land",[26] glossing over both their shortened time horizons and the impact of urbanization on surrounding countryside (3 per cent won't be the end of it). The public remains unconvinced.

I have often found, when discussing these issues, a similar reaction to the one I get on transport and climate change: if I don't like the consequences, I will doubt the evidence: the government must be exaggerating. I put this question to Neil Sinden, Director of Policy and Campaigns at the Campaign to Protect Rural England (CPRE), an organization that campaigns against greenfield building:

> "We recognize that there is a significant shortage of housing ... The housing stats demonstrate that we have been seriously undershooting the level of need, and have been for the last four or five years ... "

CPRE's differences with the government are more about types of housing and location. Second homes and empty homes cause serious problems in some areas and action could be taken to reduce both[27] but their numbers are relatively small; they do not change the basic national problem that we need a lot more housing. Any government taking office over the next decade or so will be caught between pressures to increase house building, restrain greenfield development and provide infrastructure for a rapidly growing population in a country which most people already consider overcrowded. The future of transport will depend on how governments, and voters, respond to these pressures.

Transport problems are inextricably linked to the broader structure of society. One recurring theme of the book is that there are no simple solutions; indeed the search for simple solutions is often part of the problem. That doesn't mean we have to accept things as they are. If my writing conveys a passion for some ideas and frustration with others, this book makes no claim to offer all the answers. It aims to provoke. If it prompts you to explore some of these issues further and reconsider any aspects of conventional wisdom, it will have achieved its aim.

CHAPTER 3
"There has been a war on the motorist"

Myths:
- Successive governments have been trying to "get people out of their cars".
- The cost of motoring has been rising because governments have increased taxes on fuel.
- Motorists pay for the roads through road tax.
- Speed cameras are another form of taxation – they do not reduce casualties.

Observations:
- Reducing car use and traffic at a national level was government policy between 1997 and 2000 only.
- The cost of motoring has fallen in recent years, whereas the cost of public transport has risen.
- Taxes on fuel have been falling since the fuel tax protests of 2000.
- Road tax was abolished in 1937. Roads are financed from general taxation. Vehicle excise duty is based on carbon emissions.
- There is a strong relationship between speed and road casualties.
- Serious studies – published by the RAC Foundation – show that speed cameras reduce road casualties.
- Some newspapers, such as the *Daily Mail*, have misrepresented evidence about fuel tax, speed cameras and casualties on the roads.
- Some of the commentary about the 'war on the motorist' exhibits the psychological trait of 'in-group favouritism'.

"Successive governments have been trying to get people out of their cars"

In the latter years of the last Labour government, a new phrase was used to describe government policy: 'mode neutrality' ie the Government would not try to influence whether people drove or travelled by some other means.[28] An end to 'the war on the motorist' became the policy of the Conservative-led Coalition government with a couple of announcements in 2010 and 2011. The idea that governments are actively trying to stop people driving cars is an enduring myth, believed much more widely than the petrol-headed fringe where it originated. I have also noticed it among people who would actually like to reduce their own car use, saying things like: "they want us to get out of our cars but ... " The second half of the sentence about lack of support for public transport or active travel is often well founded, but the first part seems resistant to any evidence of what governments are actually trying to do. Even many transport academics were slow to notice the changing direction of UK government policy in the early 2000s and particularly after the Conservative and Liberal Democrat Coalition came to power in 2010.

In fact, the only period in recent history when UK government was unequivocally committed to reducing car use was the brief period from 1997 to 2000, when John Prescott presided over a big department including transport, famously boasting:

> "I will have failed if in five years' time there are not ... far fewer journeys by car. It's a tall order but I urge you to hold me to it."[29]

That aspiration, enshrined in the Road Traffic Reduction Acts of 1997 and 1998 was quietly buried between January 2000, when national traffic reduction targets were abandoned, and 2002, when Prescott's big department was broken up. The white paper of 2004 no longer sought to reduce traffic, only to slow down its growth. From that point, government policy documents talked about 'reducing the need to travel by car' or supporting alternatives like buses. As we will see in Chapter 5, there is a big difference between trying to reduce something, like traffic, and trying to increase an alternative, like public transport: one does not necessarily lead to a reduction in the other.

The fuel tax protests of 2000, where road hauliers blockaded oil refineries, changed several aspects of transport policy. The strategy of using fuel tax to promote 'modal shift', from car driving to more sustainable alternatives, was reversed although hardly anyone in the media or in everyday discussion appeared to notice the repeated cuts in fuel taxes. Figure 3.1 shows the changes in the price and taxation of petrol from

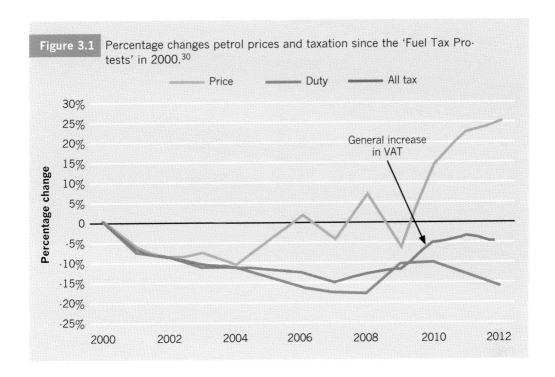

Figure 3.1 Percentage changes petrol prices and taxation since the 'Fuel Tax Protests' in 2000.[30]

2000 onwards. Prices rose after 2004 because of rising crude oil prices and a general rise in VAT. After a blip in 2008, the rate of fuel duty continued to fall.

Most taxes, like VAT and income tax, are set in percentage terms but fuel duty is levied in pence per litre. Unless the Chancellor increases fuel duty in line with inflation each year, its value falls. This distinction has allowed media outlets like the *Daily Mail* to publish misleading statements like the following in 2012:

> "the duty is still much higher than it was in March 2001, **just six months after the fuel protests which rocked Britain.**"[31]

Even after allowing for an increase in VAT on all goods in 2010, the tax on petrol was still 5 per cent lower in 2012 than in 2000.

Overall taxation of motoring in the UK was similar to other European countries in 2009.[32] In 2011, petrol tax was slightly higher than the European average (60 per cent compared to 57 per cent)[33] but other forms of taxation, particularly motorway tolls and car purchase tax, are more onerous in several other European countries.

Reduced government support for sustainable transport

The speech by Conservative transport minister Philip Hammond in 2010, declaring an end to "Labour's war on the motorist",[34] presaged the demise of the M4 bus lane and central government support for speed cameras. This was followed by changes to planning guidance on the number of parking spaces in new commercial developments (offices, retail parks etc). The previous government had introduced national maximum guidelines, which the Coalition scrapped in 2011. The most fundamental change occurred in 2013, however, with a return to large-scale road building.

For the first three years of the Coalition government (2010–13), the scale of the shift away from sustainable transport was disguised to some extent by a clever strategy of keeping a progressive Liberal Democrat junior minister, Norman Baker, in charge of sustainable transport initiatives. Baker did a pretty good job with the tiny budget he was allocated. On the same day as new £28bn road building plans were announced and the budget for high-speed rail was increased, £100m was allocated to extend the Local Sustainable Transport Fund for another year. The public, it is often said, doesn't really get the difference between millions and billions: they all sound like big numbers. If you have managed to read this far, I am sure that doesn't apply to anyone as intelligent as you, but just to illustrate the point, Figure 3.2 compares the size of the road, rail and sustainable transport announcements.

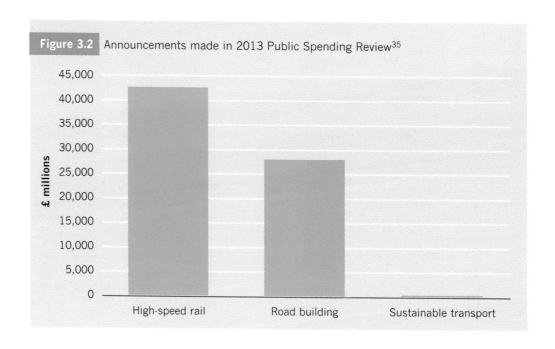

Figure 3.2 Announcements made in 2013 Public Spending Review[35]

The sustainable transport is so small it is almost invisible on this figure's scale. To put this in a broader context, Figure 3.2 shows total public spending on different elements of transport over a five-year period. 'Local public transport' is mainly support to bus services. Whereas much of the road and rail budgets are spent centrally, the vast majority of public spending on walking and cycling is provided by central government but spent by local authorities. At a local level this can create a misleading impression that cycling, in particular, is receiving a significant share of the transport cake, which it is not.[36]

The sustainability credentials of high-speed rail will be considered in Chapter 5: we are only examining the cost here. Although roads have always taken the lion's share of public spending on transport, the rail network has also consumed a lot of public money in recent years. With a few urban exceptions like Hong Kong, rail networks in developed countries always require public subsidy, but the way the British network was privatized and fragmented has resulted in poor value for money.[38] Government support to the rail industry nearly trebled in the three years following the Hatfield crash in 2000[39] and successive governments have struggled to control rail costs with relatively few improvements to the network ever since. A similar point can be made about the deregulated bus system in England outside London. Around half the revenue of private bus operators has come from the taxpayer in recent years.[40]

Governments have responded to the rail problem by allowing fares to rise faster than inflation, while private bus operators outside London are able to raise fares to

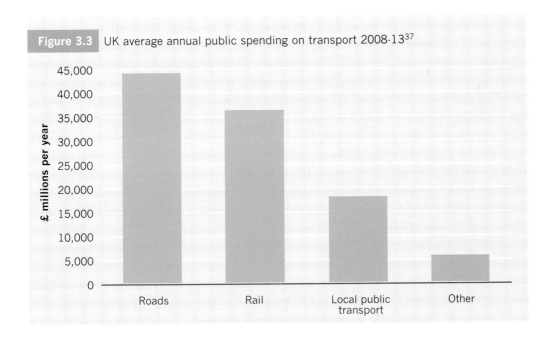

Figure 3.3 UK average annual public spending on transport 2008-13[37]

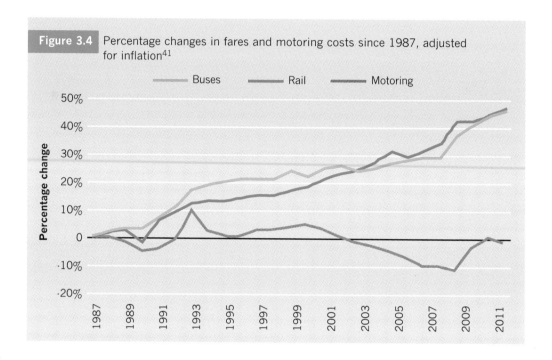

Figure 3.4 Percentage changes in fares and motoring costs since 1987, adjusted for inflation[41]

whatever level the market will bear. Figure 3.4 shows the trends in costs of different forms of travel. A combination of tax cuts and falling non-fuel costs has kept the cost of motoring roughly stable, despite the rising costs of crude oil, but bus and rail fares have both increased by nearly 50 per cent since 1987.

"Motorists pay for the roads through road tax"

The argument that motoring is overtaxed is sometimes made by comparing the revenue from taxes on motoring with spending on roads.[42] Very few taxes are ring fenced in that way, although road tax was, until 1937. Since then, roads have been funded from general taxation. As Winston Churchill argued:

> "Entertainments may be taxed; public houses may be taxed; racehorses may be taxed ... and the yield devoted to general revenue. But motorists are to be privileged for all time to have the whole yield of the tax on motors devoted to roads? Obviously this is all nonsense ... such contentions are absurd, and constitute an outrage upon the sovereignty of Parliament and on common sense."[43]

In fact, since 2000, vehicle excise duty (the correct term for what is commonly referred to as 'road tax') has been based on carbon emissions with the cleanest cars being exempt. This has not stopped either newspapers or politicians making statements like the following, from a candidate from the UK Independence party (UKIP) in Cambridge:

> "Road space is required for motorised vehicles who pay for it. It shouldn't be wasted on people who don't ... What is 'sustainable transport'? Is it using things that other people pay for?"[44]

A similar point was made in less inflammatory language by a junior transport minister in 2011.[45]

A few studies have tried to calculate whether motorists do 'pay their way', when a wider range of social costs (what economists call 'externalities') such as accidents, air pollution and carbon emissions are taken into account. These studies have generally found that the costs imposed on society as a whole exceed the tax revenues raised from motoring (the Sustainable Development Commission (SDC) presented one such study to the Coalition government in 2011,[46] who ignored the report and abolished the Commission shortly afterwards). As the next chapter will show, arguing about costs will never get us very far, whatever the calculations show. If we are interested in solving transport problems, the key issue is not whether motorists are paying enough to cover their pollution, but what are we going to do about traffic pollution, which is endangering human health in most British cities?[47]

The SDC report also touched on another favourite campaign issue of the motoring lobby, personal freedom. Most debate around urban transport looks at the freedom issue from only one side: the freedom to drive. The freedom *not* to drive, and to live without the consequences of traffic in your neighbourhood, is rarely recognised in this country. The car free developments in other European countries, described in Chapter 13, were created partly in response to a desire for that freedom.

"Speed cameras are a money-making scam"

Some of the most misleading comments by newspapers (echoed by some politicians) on the 'war on the motorist' concern speed cameras. The Coalition government's decision to scrap the national funding programme for speed cameras followed several years of campaigning by the *Daily Mail* (my favourite newspaper) which argued that the cameras were:

> "a giant 'scam' aimed at 'making buckets of money' for the government"[48]

and more recently:

"increase risk of serious or fatal crashes".[49]

The second claim was allegedly based on a report published by the RAC Foundation, from a transport statistics expert. Both that report[50] and an earlier one[51] found the cameras reduced deaths, injuries and the severity of injuries. The second report examined data from 551 cameras across 9 areas. Overall, the number of fatal or serious casualties fell by 27 per cent, but in a very few cases (21 cameras) the number of collisions had "risen markedly", hence the headline. If you have ever worked with statistics you will realize the folly of that argument: things measured in the real world go up and down for random reasons as well as underlying causes. It is entirely normal for a few 'odd cases' to buck any trend involving human behaviour.

A large body of evidence from the UK and elsewhere illustrates the simple maxim that speed kills. As assessed by the police attending the scene of a collision, breaking the speed limit, or driving too fast for the conditions, contributes to about a quarter of all deaths on the road, around 400 people a year.[52] A review of 11 studies from 6 countries using different methods produced a range of findings but all 11 found higher vehicle speeds caused a higher risk of pedestrian fatalities.[53] A study of road casualties in London between 1986 and 2006 found 20 mph zones reduced casualties by 42 per cent: the reductions were greater for younger children and serious or fatal injuries.[54]

Most people travel by different modes at different times, so 'the motorist' (half-man, half-machine), like 'the pedestrian' or 'the cyclist', is an imaginary being, necessary for the creation of myths. As driving is the most common form of travel in most parts of the developed world, research into public attitudes has shown greater empathy towards 'motorists' than to other road users. This is true not only among people who drive, but also among people who mainly travel by other means.[55] The concept of driving as being 'normal' poses one of the biggest challenges to anyone trying to change travel behaviour. It also helps to explain why the myth of the persecuted motorist is so popular.

Many of the writings and discussions about the war on the motorist display what psychologists call 'in-group favouritism'.[56] Laboratory studies and observations in real life show that people judge crimes committed by 'people like us' more leniently than crimes committed by groups we see as different from ourselves. As most people own and drive cars, there is no stereotype of 'car drivers in general', but there are stereotypes – of cyclists, of HGV drivers and of particular categories of car driver like young uninsured drivers – who can be mentally distanced from the person making the judgment.[57]

Returning to the argument that speed cameras are a revenue-raising scam: all fines raise money for the Treasury. CCTV cameras have helped to convict and fine many law-breakers for offences ranging from serious assaults to riding a pony into a branch of McDonald's.[58] Many people dislike CCTV but even organizations like Big Brother Watch that campaign against the principle apply the revenue-raising argument solely to motoring offences (subtext: committed by 'normal people like us').[59]

It is sometimes assumed that deaths on the road are caused by a tiny anti-social minority taking extreme risks. In reality, about a half of pedestrians killed by vehicles are hit at speeds of 30 mph or less (hence the benefits of 20 mph zones). The risk of killing a pedestrian rises with speed. There is a threshold at around 30 mph, above which, the risk rises more rapidly.[60] Nearly half of cars recorded in 30 mph zones are breaking the speed limit,[61] which suggests the uncomfortable observation that 'people like us' are threatening the lives of others. If you are one of those people, do you feel the urge to blame someone else? If so, you are not unusual. Whether by design or instinct, my favourite newspaper often appeals to the in-group favouritism of its readers with articles like the following, by Simon Heffer:

> "The number of motorists being caught by speed cameras will soon pass the three million mark, and the money raked off from unfortunate drivers in fines is set to soar to £180 million. **I must declare an interest. For the first time in more than 25 years of motoring, I have just acquired a speeding conviction.** I was driving at 37 mph in a 30 mph zone, down an empty main road in the middle of a clear, dry, warm night. **Call me arrogant, but like probably millions of motorists before me I felt my only crime was being caught …** "[62]

To identify with an 'in-group', people need to perceive one or more 'out-groups' who behave differently from 'people like us'.[63] The *Daily Mail* is very good at this; depending on the issue, its out-groups include: benefit claimants, single mothers, Muslims, asylum seekers and foreigners in the European Union. When it comes to transport, cyclists are often identified as the out-group. It may seem strange that so much vitriol is poured on practitioners of such a marginal activity in Britain, particularly as most cyclists are also motorists. But the significance of cycling has nothing to do with cycling itself; it lies in the contrast between the normality of driving and the abnormality of their out-group behaviour. The flip side of in-group favouritism is harsher judgment of out-groups. Compare the attitude to speeding, above, with this passage, also from the *Daily Mail*:

> "while motorists who run a red light face being filmed by camera, prosecuted and having points on their licence or worse – cyclists can 'ignore the law with impunity' said Labour peer Lord Lipsey … **He recently witnessed a pregnant woman knocked to the ground by a**

cyclist and said even MPs and peers walking near the Houses of Parliament are victims of 'aggressive' anti-social cyclists who put lives at risk."[64]

A Freedom of Information request covering 1998 to 2007 in London found no pedestrian had been killed by a cyclist jumping a red light, while 12 had been killed by cars driving through red lights.[65] I cite these two passages, not to excuse cyclists or anyone else breaking the laws of the road, but to illustrate the psychological forces that help explain why the myth of the 'war on the motorist' remains so popular, regardless of changing government policies.

It is not just newspapers which reinforce the message that driving is normal. If you have ever given up your car (as nearly half a million households in Britain do every year[66]), you will immediately begin to notice all the unconscious assumptions made by people and organizations, like the following invitation to a Christmas party organized by the staff association in the place I worked in a few years ago:

"This restaurant is near the ring road with lots of free parking, so it will be accessible to everyone."

When I pointed out that not everyone had a car and questioned the wisdom of driving to Christmas parties, I got a reply which might have come straight out of the *Daily Mail*!

All these messages exaggerate the widely held belief that all normal people drive cars. I sometimes ask my students to guess the proportion of households in Britain without a car. The ones who don't know usually underestimate; some have guessed as low as 2 per cent. In reality, 25 per cent of households do not have a car. The proportion fell for most of the period since the Second World War until 2005 – three years before the recession – since when it has stabilized.[67] Licence holding among young people, particularly young men, has been falling since the early 1990s in Britain, and similar trends have been observed across several other developed countries.[68] The average distance driven per person has also been falling slightly in Britain and elsewhere, leading some academics to write about 'peak car'. The jury is still out on the reasons for all this (there are many), but we may be seeing the start of a longer-term shift.[69] Part II will consider some of the implications of this.

CHAPTER 4
"Roads and Airports Benefit the Economy"

Myths:
- Road building boosts the economy.
- If you close a road, all the traffic is forced on to surrounding roads.
- Expanding airport capacity is essential for the economy.
- Cost–benefit analysis (CBA) enables decisions to be made on objective criteria rather than subjective value judgments.

Observations:
- Improved road access can and sometimes does suck economic activity out of some areas.
- Road building can cause economic activity to move from one place to another, but there is no sound evidence that (in developed countries) it makes any net addition to the national economy.
- Road building induces additional traffic.
- Closing roads reduces overall traffic and can benefit the economies of towns and cities.
- 74 per cent of flights from the UK are for leisure or personal purposes.
- Most of these are outward journeys leading to a net loss to the UK economy, and this loss has increased with the expansion of low-cost flights.
- All forms of CBA are simply the distillation of a series of value judgments.

"Road building boosts the economy"

"The long term benefits of road investment are well known", says a report written by the Confederation of British Industry in 2012, which argues for more road building and a partial privatization of the road network.[70] The report offers no evidence to support this claim; its authors may have felt the statement was self-evident. As economic growth has faltered in recent years, transport debates have increasingly focused on alleged economic benefits. Dubious claims about 'the economy' trump some much clearer objectives for transport such as better integration of different modes of transport, as well as reducing its impact on the environment and improving quality of life. Consequently, supporters of sustainable transport often feel compelled to express their arguments principally in economic terms. This has produced what later, saner generations may well regard as some ludicrous extremes. Several reports have attempted, for example, to justify the many benefits of walking and cycling or the disbenefits of pollution primarily on economic grounds, as though longer healthier lives could not be considered sufficient justification in their own right.

The term 'economic benefits' has three quite different meanings, which are sometimes confused in this debate:

1. Growth in Gross Domestic Product (GDP) or some other measure of economic activity is the most common. GDP may be measured at a local, regional or national level.
2. When politicians or business leaders talk of 'economic benefits' they may be referring to financial benefits for a specific group, or interests that they represent (Eg a foreword to the CBI report was written by the Chief Executive Officer of Aggregate Industries, a major supplier to road builders). These 'benefits' may or may not benefit 'the economy' as a whole.
3. Where the benefits of a project are quoted in pounds or in a ratio (like 3:1), this has usually been derived from a cost-benefit analysis (CBA), and implies a meaning quite different to the two above. CBAs are calculated in different ways, but when used by governments they usually include factors that are more social or environmental – like the health benefits from walking and cycling, or the disbenefits of pollution. The outcomes of these calculations are often described, misleadingly, as *economic* benefits. Other terms such as 'social welfare' or 'social value' are sometimes used instead. Whichever term is used, the pounds quoted are a combination of 'real' pounds and attempts to value things that are less tangible.

We will begin by considering the narrower definitions of economic benefits and return to the broader concepts of CBA later.

All public spending has a positive 'multiplier effect'. Money paid to contractors who pay their suppliers and employees will expand the economic benefits of any project, be it building roads, building hospitals or digging holes to fill them in again. The opposite is true of taxation, which reduces spending by private individuals and companies. These multiplier effects must be subtracted when trying to assess the net impact of transport infrastructure investment on the economy.

The claims made about the economic benefits of transport infrastructure, particularly road building and rail, stem mainly from the time saved by faster journeys. The estimated value of so-called 'time savings' to individuals is included in the appraisal models used by governments and transport professionals in several countries. When business leaders and politicians talk about benefits to the economy from road building or faster rail travel they are mainly concerned about the effects of time savings on businesses: additional capacity, it is argued, will reduce travel time wasted by employees, and will reduce the costs of delivering goods or materials. It will enable firms to locate in places served by new or expanded roads.

Hundreds of studies by economists, planners and transport academics have tried to test whether these claims are true. Most of these studies suffer from a problem which will recur several times in this book: the 'what causes what' problem. It is easy to show that countries or regions with more roads are richer and that road building has accompanied economic growth over time, but which was cause and which was effect?

The question is easier to answer when looking at the local effects of a specific scheme. Several studies have shown how road building changes economic activity. The effect appears to cut both ways. A new road can entice a company to open premises next to it. But it might do the opposite – persuade a local company to move away and supply a wider area from somewhere else. Both of these effects have been observed at different times and places.[71]

Several studies of the transport–economy relationship have used productivity measures, assuming that higher productivity is necessarily a good thing, and that it will make people better off. In fact, this may not always be true. The argument usually runs as follows. Productivity is generally measured in terms of output per employee (or 'unit of labour') – if output increases, firms can sell more with fewer staff, increasing their competitiveness. Across the economy as a whole, the people no longer needed will find new jobs producing additional goods or services. So the theory runs. In some circumstances this is indeed what happens, but not always. Consider the effect of a new road on shops in a small town or village. A quicker journey to an out-of-town hypermarket will transfer spending from the local shops to the hypermarket. This will increase 'productivity', because hypermarkets employ fewer staff per unit of sales than local shops. But the effects on the local economy and on overall employment may be negative. If the people who are no longer employed by local shops remain unemployed, then the overall effect on economic well-being will also be negative. This effect on retail has also been observed in practice.

So does building roads or other transport infrastructure merely move economic activity from one place to another, or does it truly expand the overall economy? Several studies have tried to address that question: some suggest that it does, some that it does not, but none have provided a really convincing answer.[72]

One American study found quite a clever way to address the problem of 'what causes what?' It used measures of highway spending and industrial output in 55 counties of California over a 19-year period, and also looked at the effects on neighbouring counties.[73] It seems that highway spending in each county tends to have two effects:

1. It increases economic activity in that county.
2. It reduces economic activity in neighbouring counties.

Interestingly, the positive effect of spending in each county was slightly smaller than the negative effect of spending by their neighbours. The author cautions readers "not to read too much into the numbers". He prefers to believe that highway spending causes national or state-wide growth, although his own findings suggest the opposite to be true.

Ultimately, this argument is probably impossible to prove one way or the other. One study of international data did suggest another plausible conclusion, however, which is consistent with findings in other related fields.[74] In developing countries, road spending does seem to benefit national economies: this is particularly likely in countries that do not have a network of paved roads covering the whole country. More developed countries may have achieved one-off economic gains in the recent past, from building a motorway network, for example, but those gains cannot be repeated by continuing to expand road capacity.

From a local or regional perspective, there is a growing body of evidence for economic benefits from the closure or demolition of roads, where this improves the quality of the urban environment. The most dramatic example of this occurred in Seoul, South Korea, where an elevated motorway in the city centre was removed creating a linear park along the banks of a stream, which the motorway used to cover. New cycling and walking routes and an expanded busway network replaced some of the lost road capacity. Both residential and commercial property values in the area rose substantially as a result.[75] Similar effects have been observed in the UK from pedestrianization and other improvements to urban environments resulting from reduction or removal of traffic.[76]

It is often assumed that when a road is closed, all the traffic is forced on to surrounding roads. Many of the computer models used by transport planners make that assumption, which is one reason why highway authorities often resist proposals to reduce road capacity. However, the evidence suggests this assumption is wrong: part of the removed traffic simply disappears.[77] There are several reasons for this. Some people switch to walking, cycling or public transport. Some people decide not to make some trips; they may find what they want locally or put more thought into 'trip chaining' (ie arranging to do several things in one trip). The relative importance of each of these is likely to vary from place to place; evidence on this is still rather limited.

The opposite effect has also been demonstrated across a range of different circumstances in the UK[78] and elsewhere:[79] increasing road capacity by building or widening roads *induces* more traffic. The reasons for this include: extra trips, longer trips, reduced trip chaining, and in the longer term relocation by individuals and firms who decide they can live or base themselves further away.

"Expanding airport capacity is essential for the economy"

The case for expanding airports also rests largely on claims about economic benefits. Many consultants and a few academics have tried to find evidence to support or refute these claims. Those who are paid by the aviation industry generally claim to find what they set out to find,[80] but all of them have either encountered or ignored problems similar to those surrounding the road building debate. There is a strong relationship between air traffic and economic activity, but what causes what?

Looking at national data for 48 countries over 7 years, the International Air Transport Association (IATA) found a relationship between labour productivity and a measure of connectivity by air.[81] As with the earlier discussion about roads, there is a strong positive relationship at lower income levels. The poorest countries have very low connectivity and labour productivity; both measures are much higher in richer countries. But if we compare the countries of North America and Western Europe to each other, there is no discernible relationship between the two measures.

For the UK there is another important factor. Around 81 per cent of the international flights from UK airports are for leisure or personal reasons. Of these, flights by UK residents outnumber flights by overseas residents by a factor of nearly 3-to-1 as illustrated in Figure 4.1.

As a result, outbound British tourists spend considerably more overseas than incoming tourists spend in the UK. This 'tourism gap' widened dramatically with the growth of low-cost flights around the turn of the century. The recession and growth

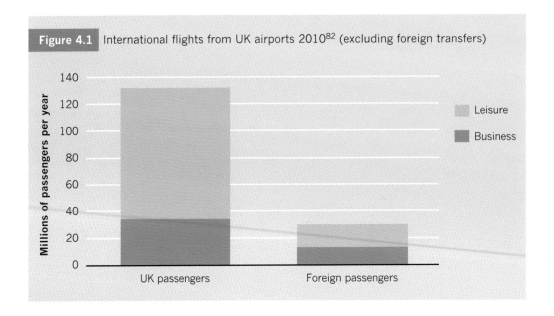

Figure 4.1 International flights from UK airports 2010[82] (excluding foreign transfers)

in 'staycationing' reduced the gap slightly, but it still represented a net loss of around 1 per cent of GDP in 2012, roughly equal to the direct contribution of the aviation industry to the UK economy.[83]

Reports written for the aviation industry or by government[84] either ignore the tourism gap or claim that limiting aviation capacity would not improve the situation. The issue is not straightforward. It critically depends on how British residents would react to a rise in the price of flying, in the short and longer terms. The effects are likely to include: an increase in domestic tourism, an increase in shorter-distance tourism using ferries and the Channel tunnel, a possible reduction in overall travel and an increase in spending on other things.

An increase in the use of ferries would also stimulate the economies of ports, some of which have suffered from the growth of low-cost flights, with several ferry links disappearing in recent years. This raises another issue often overlooked in such discussions. The distribution of economic activity may be just as important, or more important, than any overall effect. A policy of air travel restraint would stimulate ferry ports and British holiday destinations. A policy of expansion would stimulate the economy around the main airports, particularly in the south-east. Which of those two would we prefer to see? Would we accept some trade-off between overall GDP and a 'better' distribution? No calculation can answer those questions: they require a value judgment.

"Cost–benefit analysis is an objective way of making decisions"

That leads us on to issues raised by cost–benefit analysis (CBA). The modified form of CBA used by the UK Department for Transport (DfT) to appraise transport projects allows for the exercise of some judgment. Its guidance on transport appraisal (also used by local authorities and other public bodies) allows for some environmental and social disbenefits, such as noise and air pollution, to be 'monetized' so that they appear as a cost to the project. Other impacts such as landscape and biodiversity are summarized in a separate table, but are not monetized. The total of the monetized benefits is the figure which is usually quoted by politicians when they talk about the economic benefits of a transport project. As the monetized elements seem to carry more weight, some opponents of airport expansion have argued that a wider range of disbenefits should be monetized. If all these elements are included, airport expansion yields a negative benefit-to-cost ratio, some studies have found.[85]

For most projects, the monetized benefits mainly come from the value of time saved through faster travel. Time savings to businesses are estimated using average wage rates for different types of traveller: time savings for car drivers are valued more highly than for bus passengers, for example. Time savings for non-work travel are based on estimates of how much people are willing to pay for a quicker journey: these amounts are generally much lower than wage rates, so the business time savings are often the biggest 'benefit' in a CBA for roads or high-speed rail.

Many people have pointed to flaws in these methods. Some of my colleagues at UWE have studied how business travellers use their time: much of it, particularly when travelling by train, is spent working.[86] This suggests that the value of business time saved by faster travel is routinely overestimated.

There is a more fundamental problem with the whole concept of CBA. CBA assumes that the benefit or cost to society of a project or policy is equal to the sum of the benefits and costs to individuals and companies. Those benefits may be measured in direct financial terms, by indirect measures such as property prices, or by some notional willingness to pay to gain something (such as a faster journey) or to avoid something (such as air pollution). Just because individuals value something in a certain way, why does society have to agree with them? In practice, there are many circumstances where most people would disagree with that view.

Consider the relationship between travel time, driving speed and road safety. Many transport projects have the effect of reducing speeds and increasing travel time. Should transport planners considering a major project that includes traffic calming be subject to a CBA that compares the value of the 'time lost from slower driving' to the value of the 'lives expected to be saved'? The DfT guidance values human lives in the same way as it values travel time. It is based on estimates of what people would be willing to pay to avoid a death (someone else's rather than their own!).[87]

Consider two roads where speed cameras might be installed. On our 'Fast Road' drivers are driving much too quickly for the conditions, but not on our 'Slow Road'. Fast Road has more speed-related deaths. Using CBA, the benefits would come from the value of 'lives saved' and the costs would come from the value of the 'time lost'. Speed cameras would save more lives on Fast Road but would also slow down more speeding traffic, causing greater loss of travel time. So depending on the balance between the two measures, the cameras could save more lives on Fast Road but generate a higher benefit-to-cost ratio on Slow Road. More deaths would be needed to justify speed cameras on Fast Road than on Slow Road; hence CBA makes it easier to justify speed reduction measures in places where they are not needed.

The pure application of CBA produces many absurd or ethically dubious outcomes,[88] so in practice there is often pressure to change the rules or simply ignore the CBA. Consider the issue of a person's earnings and how to value their time. In the pure form of CBA, the time of wealthier people would be worth more and the 'willingness to pay' (to reduce air pollution, for example) would be greater in richer parts of the country. To avoid this problem, the DfT guidance uses average national earnings and average willingness to pay. This could be seen as changing the rules because we do not like the outcome, but it also calls into question the fundamental principle underlying CBA. Are time savings for wealthy people really worth more to society than time savings for the poor? How can we answer this question without introducing value judgments?

Supporters of CBA argue that it provides an objective decision-making process, which can be separated from short-term political goals.[89] Although its guidance allows for the exercise of some discretion, the DfT states that it will only fund projects where the monetized benefits are greater than the monetized costs because:

> "to do otherwise would be supporting proposals which diminish overall social value".[90]

It should be apparent from this discussion that whatever system of CBA is used, the outcome is the distillation of a series of value judgments. It would be impossible to say whether the outcome of a particular calculation was 'right' or 'wrong' since every element of the calculation is based on a value judgment.

The requirement to appraise each publicly funded project individually makes strategic decision-making more difficult. The benefits of integrated transport networks are nearly always greater than the sum of their individual parts. Many of these benefits become apparent only when the final pieces of a network are joined up. This is particularly true of urban rail / light rail and cycling networks. To allow for these benefits when appraising the earlier phases of a network would require assumptions about the timing and availability of funds for all the later phases. When the assumptions of each project are separately scrutinized, it is not surprising that British authorities and transport planners have tended to focus on short-term ad hoc measures rather than coherent long-term strategies.

The weight attached to travel time savings has favoured road building and made many other options difficult or impossible. Projects that involve removing or reallocating road space to improve the urban environment or the reliability of public transport are often hampered by the fact that additional journey time for motorists is counted as a cost. Whether this case-by-case focus has succeeded in saving travel time in the longer-term is debatable. The DfT claims the English appraisal framework "is seen as a leading model of documentation and has led the way in applying CBA more widely"[91] prompting one transport consultant to write in *Local Transport Today*: "if our transport appraisal methods are so smart, why is our transport system so poor?"[92]

Attempts to incorporate climate change into CBA through carbon pricing reveal its most serious weakness. Unlike the 'cap and reduce' principle used in the European Carbon Trading System, allocating a price for carbon enables companies and individuals in the present generation to collectively pay more to emit more. The long-term disbenefits may be weighed against short-term preferences for more travel or quicker journeys. If the outcome of those decisions produces catastrophic climate change, what price could possibly justify such decisions?

Transport economists and many academics obsess about calculation methods, whether the right factors have been included, whether the CBAs used to justify the high-speed rail network HS2 were right or wrong and so on. All of these arguments miss the point. Where elements of a calculation are based on opinions, the outcome will also be an opinion. The idea that getting the calculations right could show what ought to be done is an example of the calculation fallacy. As it is currently used in transport decision-making, this makes strategic decision-making more difficult, and is particularly ill-adapted to deal with the challenges of climate change. What might replace CBA and how we can make better decisions will be considered in Chapter 16.

CHAPTER 5
"All we need is better public transport"

Myths:
- The main solution to our transport problems is better, more frequent, cheaper (or free) public transport. If public transport is good enough, people will use it, instead of driving.
- "The commercial model [for running buses] outside London has delivered all of the benefits and at a fraction of the cost to taxpayers." (Stagecoach plc)[93]
- "In contracting private companies to operate trains, Government has devised a winning formula where competition injects new ideas and skills to enhance services, attract more passengers and keep costs down." (ATOC)[94]
- "We will establish a high-speed rail network as part of our programme ... for creating a low carbon economy." (DfT)[95]

Observations:
- Public transport improvements *on their own* make little difference to the habits of car drivers: they are more likely to generate additional trips, or to reduce walking or cycling.
- Based on typical behaviour changes, doubling bus use across the UK would reduce car driving by around 1.3 per cent.
- The government's own projections show £42.6bn spent on high-speed rail will reduce national car driving by 0.02 per cent.
- Even these may be overestimates: removing traffic from congested roads during peak periods frees up space for others to drive, unless constraining measures are also taken.
- Fares have increased and patronage reduced under the deregulated bus system in England outside London.
- Publicly owned rail franchises have performed better than private competitors.

"Better and cheaper public transport will solve our transport problems"

The belief that better public transport will solve all our transport problems is understandable. However, while public transport improvements are an essential *part* of a sustainable transport strategy, *on their own* their potential to reduce traffic and emissions is limited. There are several reasons for this. One of them is a question of scale.

Several cities, and some smaller towns, have shown how the use of public transport can be substantially increased over a number of years (some of these will be discussed in Chapters 12 and 14). The best example in Britain is London, where a big investment in the bus network, among other measures (including traffic restraint), doubled the number of passenger trips taken between 1994 and 2011. Although the underground's capacity was more constrained, trips on it also increased by more than half over the same period. Together, these contributed to a fall in the share of driving from 32 per cent to 23 per cent over the same period.[96]

To achieve anything like this scale of change at a national level would be far more challenging. The national pattern of transport movements in Great Britain is shown in Figure 5.1.

Cars and vans account for six times the distance covered by public transport. If public transport use could be doubled as a direct result of a reduction in car use, the distance covered by cars would fall by one sixth. But how many car journeys would be replaced in reality?

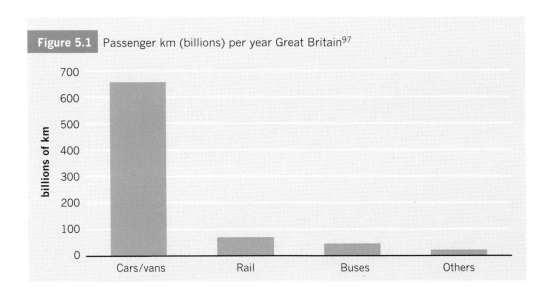

Figure 5.1 Passenger km (billions) per year Great Britain[97]

Buses, trams and trains attract new passengers for several different reasons. The first two columns in Table 5.1 show the impacts of a tram network (in Manchester) and a bus rapid transit (BRT) network (the Kent Fastlink, which runs partly on separate busways). Passengers were asked how they had previously made that journey. The final column shows a projection by the DfT of where the customers of the planned HS2 high-speed rail network would come from. The circumstances and methods of each study were different, but all illustrate the key point: that substitution of car (or air) journeys forms only a small part (between 8 per cent and 21 per cent) of the total effect. Much more of the effect is due to passengers making new journeys, transferring from other forms of public transport, or from walking or cycling.

Table 5.1 The effects of new public transport networks			
Previous journey	Manchester Metrolink (passenger survey)[98]	Kent Fastrack Bus (passenger survey)[99]	HS2 (projection)[100]
Car	21%	19%	8%
Train	32%		65%
Air			3%
Bus	20%	60%	
Walk/cycle		11%	
Other	1%	4%	
New journey	26%	6%	24%

Manchester's experience illustrates both the potential and the limitations of investment in public transport. Opened in 1992, the first phase of the Metrolink tram network (Figure 5.2) exceeded the original passenger forecasts. Like many tram systems it has been very popular, suffering from overcrowding at some peak times. It has been expanded and extended in several directions since then, but its impact on the overall travel behaviour of the city has been modest. The 2011 census shows 1.4 per cent of Manchester's working population used the tram as their main method of transport when commuting – and only a small proportion of those would otherwise have driven by car or van.

Many more people in Manchester (23.5 per cent) commute by bus than by tram, as the bus network is more extensive: far more people live within walking distance of a bus stop than a tram station. This is true of most cities with tram or local rail networks, which is one reason why successive governments have emphasized – and arguably overemphasized – the potential of the bus to change travel behaviour in urban areas. Unfortunately, the conventional bus, running on streets as opposed to separate busways, is the least attractive of all public transport modes to car drivers. Another study

Figure 5.2   Manchester Metrolink

in Manchester, of recently improved bus services, showed only 10-14 per cent of passengers had a car available, which they could have driven instead of taking the bus.

Conventional buses are slow, typically averaging around 10 mph compared to around 20 mph for stopping trains or underground networks, with trams, BRT and urban car driving somewhere between the two. Most bus services follow radial routes, in and out of town or city centres, where the largest concentrations of journey destinations are to be found. The 'largest concentration' does not mean 'most destinations', however. Although more journeys are made from suburb to suburb, the pattern of these journeys is generally too dispersed to support viable bus services. Orbital bus services, avoiding city centres, often require a public subsidy. Most of the journeys currently made by car would require at least one change of bus. But travellers are more resistant to bus-to-bus changes than to changes between other modes of transport: if the alternative is a two-bus journey, most people with a car available will drive. All of this means that expanding bus services is less likely to reduce the number of journeys taken by car and more likely to encourage extra journeys or transfers from journeys usually made by walking or cycling. This is particularly true of free bus travel. The town of Hasselt in Belgium offered free bus travel in 1996, and also expanded the number of bus lines. Starting from a low base, bus use increased tenfold, but two-thirds of the extra journeys were made by people who were already bus users; 21 per cent switched from walking or cycling; only 16 per cent switched from driving. The largest study of concessionary pass holders in Britain found a similar pattern: only 18 per cent would otherwise have driven.

Looking back at Figure 5.1, if bus use in Britain could be doubled – a pretty challenging objective – and if we assumed that 20 per cent of the extra journeys replaced a car trip, this would reduce national car mileage by just 1.3 per cent. But if bus improvements were the only changes on offer, even that would be an over estimate. The radial urban roads most heavily used by buses are often congested at peak times. Congestion 'finds its own level' to some extent: congested roads persuade some people to change their mode of travel, or reduce the number or distance of journeys. Anything that removes cars from congested roads frees up more space for other people to drive, unless measures are taken to prevent this. So the 'carrot' of public transport improvement without the 'stick' of traffic restraint is unlikely to make much difference at peak times.

This is one of several reasons why park and ride schemes do not, generally, reduce traffic volumes.[101] The other reasons are: that some people drive to a park and ride instead of taking a bus; some people make a detour to drive there; and, like all transport offerings, park and ride services create opportunities for extra journeys.

Cities with relatively high levels of bus use where buses converge on a single centre, like Oxford, have encountered a further problem: most of the local air pollution (especially particulates, tiny pieces of soot which damage the lungs) is caused by diesel buses. In Oxford, a low emission zone is expected to reduce but not to solve the problem (and unfortunately, low emission buses consume more fuel and emit more carbon than ordinary buses).

Like the petrol-fuelled car, we are probably stuck with the diesel bus for quite a few years to come. Among many shortcomings, its average carbon footprint per person is similar to that of a small car servicing only 1 person, based on average occupancy levels, which are currently around 16 passengers in London and 9 elsewhere. Filling those empty seats is an obvious goal but not easy to achieve. The pattern of demand for bus use is highly variable, and unlike (some) trains, the size of a bus cannot be varied in response to fluctuating demand. This leaves two other possibilities: varying frequency and varying fares. But regular services and fixed fares are more likely to attract regular customers than services which are expensive at peak times and less frequent at other times. There is also a need to provide an essential minimum service in more sparsely-populated areas. Some compromise between those objectives is unavoidable, so some buses will always run with few people on them.

CHAPTER 5 All we need is better public transport

Figure 5.3 Most air pollution in Oxford city centre is caused by buses

In 1986, the Thatcher government deregulated bus services everywhere in Britain except London: any operator 'of good repute' can set up, change or close down a bus service providing they register the service or the change with the traffic commissioners. Deregulation was followed by a rapid and sustained fall in bus use outside London, as illustrated in Figure 5.4. Free concessionary fares introduced in 2006 (2002 in Scotland) temporarily reversed the trend, at considerable expense to the taxpayer and the dwindling number of full fare payers, who ended up paying even more.[102] Outside London, peak-time commuting by bus has declined even more than shown in Figure 5.4.[103]

The system in Britain outside London is an international odd case. Most other European countries organise their buses like London: the authorities decide on services, fares and ticketing methods, and invite operators to bid for contracts to run them. Since the early 1990s, bus use has fallen in most European countries as car ownership has risen but the decline in Britain has been much more rapid than most.[104]

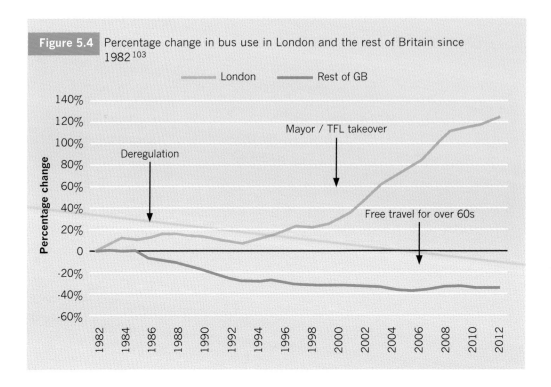

Figure 5.4 Percentage change in bus use in London and the rest of Britain since 1982 [103]

Interestingly, fare increases are not the main reason for the difference between London and the rest of Britain shown in Figure 5.4. London was compelled to raise its bus fares under the Thatcher government, and over the period as a whole, fares increased more rapidly in London than elsewhere.[105] Fare levels and value for money are not necessarily the same thing. The simplicity and flexibility of fares in London are the biggest advantage over the system in the rest of the country, where years of talk about integrated ticketing has failed to produce anything to compare with the 'Oyster card', the smartcard used on all forms of public transport in London. Smartcards are slowly being introduced in other parts of Britain, but with more ifs, buts and exceptions than the Oyster card. In trying to explain this situation, *Traffic Jam*, a book that reviewed British transport performance a few years ago, provided a wonderful piece of academic understatement:

> "The lack of political appetite to re-regulate, despite the rude health of the bus network in London and the successes of other European networks, immediately raises the question as to whether vested interests are at work."[106]

The book goes on to describe payments to political parties, who then changed their policies and bus companies paying for actors in dinosaur costumes to pose for the media outside the Houses of Parliament when regulation of buses was debated. The Labour government that took office in 1997 and set out to rectify some of the problems caused by deregulation. It introduced powers for councils to re-regulate their local bus services but for over a decade no councils dared use them. Those who seriously thought about it were lobbied, and if that didn't work, were threatened with legal action. Some operators even threatened to take action under the European Declaration of Human Rights, arguing that regulation (which is normal elsewhere in Europe) would threaten the 'peaceful enjoyment of their possessions' – for which they would demand compensation.[107] A few years ago the Competition Commission investigated the workings of the bus industry. Its interim report appeared to favour some form of re-regulation.[108] Ferocious lobbying[109] was followed by an abrupt about-face in the final report.[110]

In the early days following bus privatization, some operators instigated 'bus wars', flooding the streets with cheaper or free bus services to drive smaller competitors out of business. One notorious example occurred in Darlington in 1994, where Stagecoach ran free buses on the routes of the Darlington Transport Company (DTC) driving its competitor out of business.[111] The Monopolies and Mergers Commission launched an inquiry after DTC had ceased trading, concluding that Stagecoach's actions were "predatory, deplorable and against the public interest" but that it had no powers to impose any penalties.[112]

Once the big companies had established local monopolies, the need for such tactics receded, but similar forces are still at work in a more subtle way today. A bus industry insider agreed to talk to me "off the record" (and names have been removed to protect his anonymity):

> "There is a culture of: 'this is my patch, that's your patch … ' **Does competition raise standards? I used to think so but I'm now not so sure. It depends how good or bad that competition is.** Company A have come into this area, started eating away at Company B's (one of the big five operators) revenue sources. **This has prompted Company B to a complete re think …** They decided they need to concentrate on two seemingly opposite things: core corridors where they can see organic growth, but also they need to eliminate other operators who are generally on concessionary contracts from local authorities [for unprofitable routes, late evenings and weekends]. **The 1986 Act says local authorities cannot compete with commercial operators.**
>
> If you look at the VOSA [government agency] website, you will see the registrations Company B have put through. **Some of those are**

almost identical to the routes run by Company A for the local authority, which will in due course withdraw them ... Once they have eliminated the competition, you could ask: 'Is it realistic that they can operate a service commercially with twice the frequency and twice the costs that a local authority was operating at a loss?'"

Across the UK, deregulation of buses has led to decline (with a few exceptions, such as Brighton, discussed in Chapter 15). I ask my industry insider whether he believes the London model would improve the situation if it were rolled out nationwide:

"If you are looking at the London model without paying higher taxes, I think it would run like it does in Northern Ireland. It looks very much like most bus services looked in 1985. In fact, I think some of the buses date back to 1985!"

"Private operation of trains is a winning formula"

There are some similarities and some differences between the bus and rail industries. The rail industry is more heavily regulated, and rail use has been increasing. Many of the UK's main rail lines are now at or reaching capacity: new investment in the rail network is now essential, but likely to be expensive. As described in Chapter 3, the form of privatization (where the industry was broken into many small franchises) has proved expensive for taxpayers and passengers. Like the bus industry, it has created an oligopoly, which controls the market and influences the politicians who make the rules.

The structure of the rail industry has created a one-way bet for the few big companies that control it (rather like the big players in the banking industry). Franchises are often let on the basis that the government will pay the operators in the early years, with the payments reducing or reversing in the later years. The operators know it would be political suicide for a government to allow rail services to stop running, so if the later years prove less profitable, they can simply hand the franchise back to the government. This has happened three times in slightly different circumstances:

1. In the early 2000s, the Southeastern franchise was run by French company Connex, which was failing in both service and financial terms. The Strategic Rail Authority (SRA) initially bailed it out with a one-off £58m payment in addition to the contractual £55m subsidy.[113] When that proved insufficient, the SRA effectively renationalized the franchise in 2003.[114] Financial performance improved, passenger complaints fell and punctuality began to outperform the privately run franchises elsewhere in southern England. Concerned that this might strengthen

CHAPTER 5 All we need is better public transport

calls for renationalization, ministers played down the success of Southeastern, and instructed the SRA to do likewise[115] before returning the franchise to the private sector in 2006.

2. The East Coast franchise has twice run into financial difficulties: in 2006, when the franchise was removed from GNER, and in 2009, when the new operator, National Express, handed the franchise back to the DfT. On the second occasion, the franchise was taken back into public ownership. Again, the public operator proved an embarrassment to ministers by improving passenger satisfaction and achieving the lowest level of public subsidy among all 15 rail franchises: 1 per cent compared to an average of 32 per cent.[116] Again, ministers decided to return the franchise to the private sector, from 2015.

3. First Group, who won the newly created Greater Western franchise in 2005, took the profits in the early years then invoked a break clause to hand back the franchise in 2013 before it was required to start repaying larger sums to the Treasury.[117] First Group was allowed to bid again for the new franchise, and was shortlisted. Following a mix-up by DfT officials, the re-tendering of several franchises including Greater Western was put on hold in 2012, leaving First Group to continue running the trains to the west under a 'negotiated short-term franchise' in the meantime.[118]

Rail industry publicists claim that rising use of the railways is due to privatisation and franchising,[119] but the performance of the few publicly owned franchises calls that into doubt. Demographic changes and congestion on the roads are more likely explanations, and in any case the rise began over a decade before privatization.[120]

Better value for money is another common argument in favour of privatisation, which would be difficult to apply to the railways. The size of public subsidy to the private rail operators has been disguised to some extent by the growing debts incurred by Network Rail. The track charges paid by the train operators to Network Rail are insufficient to cover the cost of maintaining the tracks, stations and other infrastructure. Government grants make up only part of the difference, forcing Network Rail to borrow the rest. These debts now equal over £500 for every person in the country[121] and will have to be honoured by the taxpayer at some point.

"High speed rail will help create a low-carbon economy"

As a means of reducing traffic or energy consumption, high-speed rail represents spectacularly bad value for money. For a budgeted cost of £42.6bn, the projected reductions in journeys shown in Table 5.1 would represent around 0.02 per cent of national car journeys and 1.5 per cent of flights.[122] The DfT's *Economic case for HS2* [8] says: "There remains a significant degree of uncertainty on the impact of HS2 on the actual carbon emissions of transport in the UK." High-speed rail is the most energy-intensive form of public transport. Whether it increases or reduces emissions will depend on future changes to electricity generation. One thing is certain: modal shift on that scale will make very little difference to national carbon emissions. HS2 will help to relieve capacity problems on the main north–south lines, though also at great cost, which is likely to worsen the financial constraints on the rest of the network.

This chapter has challenged the view that public transport improvements can substantially reduce traffic and pollution on their own. It has also argued that the current structures of the bus and rail industries present problems to any government trying to shift travel from driving to public transport. None of this is intended to deny the importance of public transport, nor the possibility of modal shift in the future. Chapters 12 to 15 will look at how public transport has contributed to more sustainable patterns of movement in several cities in Britain and the rest of Europe.

CHAPTER 6

"Car ownership isn't a problem – only car use"

Myths:
- Car ownership is not a problem, only car use. We know this because Germany has higher car ownership but lower car use.
- There's no point in trying to limit residential parking – people will always find a way to own and drive cars.

Observations:
- Car ownership is the strongest predictor of car use, at the household, neighbourhood and national levels: people who don't own cars rarely drive.
- Germans own more cars and drive them more than Britons.
- Parking limitations can and sometimes do reduce car ownership.

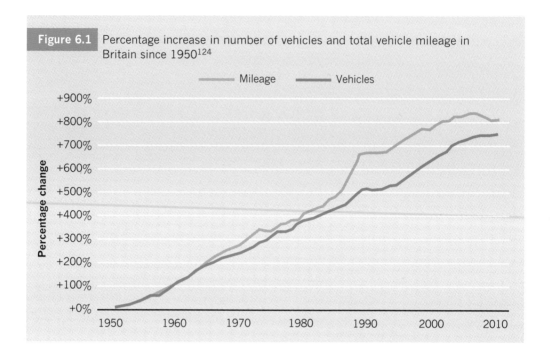

Figure 6.1 Percentage increase in number of vehicles and total vehicle mileage in Britain since 1950[124]

I have lost count of the number of times I have heard transport professionals, civil servants or politicians arguing for these myths, or variations on them. When challenged, the Germany myth is often wheeled out as justification. Like the myths in the previous chapter, they serve a useful political purpose: to convince people that existing policies can make a big difference, and that difficult political decisions can be avoided.

Whatever the cause and effect, studies from across the world have shown that car ownership is a strong predictor of car use – often stronger than any other factor. Put simply: the more cars people own, the more they drive. This is true at the individual, the household, neighbourhood and national levels over time.[123]

Figure 6.1 shows the relationship between the number of registered vehicles and the distance those vehicles are driven each year, since 1950. The total mileage increased faster during the Thatcher years as government policy explicitly encouraged car driving, particularly through road building, and has decreased in recent years as the growth in car ownership has also slowed. Whatever the reasons for these changes, Figure 6.1 shows a pattern of relatively minor divergence around two strongly related trends.

The origins of the myth about Germany can be traced back to a report published in 2000 by the now defunct Commission for Integrated Transport (CfIT).[125] The report was rather brief, based on national statistics and general observations rather than any systematic analysis. It pointed out that car ownership was higher in Germany than in the UK, but claimed that car use was lower. "High car ownership" it argued "is no barrier to lower car use." Several of the report's observations, about differences in transport policies, were relevant and did help to focus attention on good practice in German cities. In its early years, before ministers lost interest in integrating transport, the CfIT did a useful job in advising and challenging governments. It is unfortunate that one of its few lasting legacies has come from one misleading headline, which has influenced conventional transport thinking ever since.

International statistical comparisons should always be treated with some caution: definitions and data-collection methods vary and the published information may not always reveal all the differences. There are also several different ways of measuring 'car use': based on trips or on distance and per person or per vehicle. The CfIT report used average passenger km-per-person-per-year. On that basis, at that time car use in the UK was 12 per cent higher than Germany whereas car ownership *per household* was 21 per cent lower. As the report explained, those differences were partly due to smaller household sizes in Germany. To understand the reason for this, consider the difference between couples and households with one person. Overall, couples will own more cars than households with one person but not twice as many (because some couples will share a car). So comparing two similar countries, the one with more single people is likely to own more cars. Those cars will not be driven

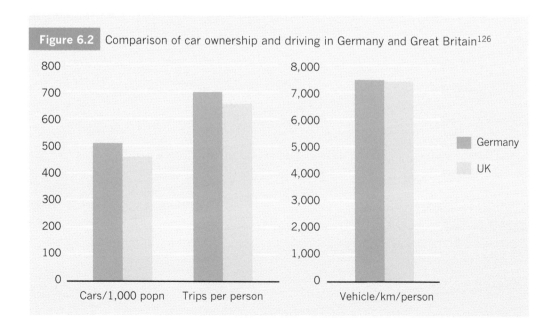

Figure 6.2 Comparison of car ownership and driving in Germany and Great Britain[126]

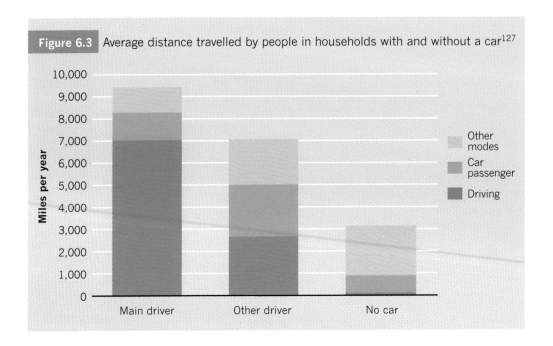

Figure 6.3 Average distance travelled by people in households with and without a car[127]

quite as often, or quite as far, as cars owned by couples. But the country with more single people (Germany) will have more cars with one person on its roads. In fact, the volume of traffic *per person* in Britain and Germany was quite similar, even then.

Since then car driving has been falling in Britain whereas it has continued to rise in Germany. By the end of 2010, Germans owned more cars and drove them more than Britons, on several different measures (see Figure 6.2).

Looking across Europe, the relatively small differences *between* countries are dwarfed by much bigger differences *within* countries. So how does car ownership affect car use within a country? Figure 6.3, drawn from the UK National Travel Survey, provides a few clues. The 'other drivers' are mainly people in one-car households where someone else uses the car more than they do. Households with two or more cars may have more than one 'main driver'.

The figure shows that people with cars make more trips and travel further. Most adults without cars do have driving licences: they may occasionally borrow or hire a car; a few belong to car clubs. (Car clubs are membership organizations that provide hire cars on flexible terms to their members. In other English-speaking countries this is usually called 'car sharing', which confusingly has a different meaning in Britain!) But overall, this graph confirms what we would expect: adults without cars don't travel much by car, and they very rarely drive themselves. If this seems like a statement of the blindingly obvious, why do so many intelligent people try to argue that car ownership makes no difference to car use?

We will return to that question later. In the meantime, if we ask "What causes what?", the picture becomes more complicated. Income is an important influence on both car ownership and use, but the relationship is not as straightforward as it used to be. Over recent decades, people with low incomes have acquired more cars, as the cost of motoring has fallen and public transport has become more expensive. A quarter of households, containing about one in six adults, do not have a car.[128] These proportions have been roughly stable since 2005. When asked why they don't own a car, only a minority of these people mention the cost: 32 per cent in one survey,[129] 44 per cent in another (using slightly different definitions).[130] The most common explanations apart from cost suggest that some people don't feel they need one.

Most of these people live towards the inner areas of larger cities. But you will find some of them everywhere: in small towns, villages, even a few in remote rural areas.[131] Stable proportions do not mean stable populations: every year some people acquire cars and other people give them up. Many of these changes occur at moments of life transition, like moving house, changing jobs, having children, divorce, retirement or when an old car fails and the owner decides: actually, I could do without one.

Some studies have tried to probe what people mean when they say 'I need' or 'I don't need a car'. A picture emerges of changing circumstances and choices that create or remove the need for a car. A few years ago I interviewed people who had either got rid of a car or bought one again after living without one. Giving up a car usually followed a move to a more accessible location. For some, this was a personal preference, such as one woman living in Glasgow:

> "I live and work in a situation where I don't need to drive. **I've done that on purpose.** I've made choices so that I don't have to drive."[131]

For others it happened more by chance. One man explained that he gave up a car when he started university, mainly to save money. He always imagined he would buy another one when he started work, but then he moved to a Midlands town where he found he didn't need one:

> "There was no real thing that said to me: right I must get a car. 'Cos I'd got so used to using a bike I was actually quite happy with how I was doing it … "

For those who lived in smaller towns or suburban locations, access to good rail services was important. They would cycle or use buses for local travel, but several of

Figure 6.4 Poole Quarter – car parking

Figure 6.5 Poole Quarter – boundary road

them explained that they needed access to a well-connected railway station for their longer-distance travel.

Part of this study examined how parking limitations in a 'low car development' influenced car ownership and use. Poole Quarter was built in the late 2000s, within walking distance of the town centre in Poole, Dorset. It is a fairly high-density development, with one parking space for each house or flat in most cases (Figure 6.4). The surrounding streets (Figure 6.5) had double yellow lines and some pay and display parking for visitors.

Just over a quarter of the residents had reduced their car ownership when moving there. Nearly half said they were driving less than they used to. This was partly due to location – walking distance of the railway and bus stations – but also because some people changed from being 'main drivers' to 'other drivers' or 'non-drivers', using the terminology above. So can we conclude the parking limitations *caused* people to drive less, or did these people all choose to move somewhere more accessible that enabled them to do what they had already intended?

Individual interviews revealed examples of both. Some people chose to move there, after retirement or divorce for example, partly because they wanted to be less dependent on a car. Other people moved there for completely different reasons and reduced their driving afterwards.

This evidence sheds some light on another aspect of the 'what causes what?' problem: the question of self-selection. We know residents of some places behave differently from residents of others: people in suburbs drive more than people in city centres, for example. This is true even after allowing for factors like income and household type. But does the place cause the behaviour, or does it just attract different types of people? This has been a long-running controversy in transport research and though these things are difficult to measure, we now know that both effects occur.[132] It is not just a question of choice: people's preferences are partly influenced by where they live and the behaviour of those around them.

"There's no point in trying to limit residential parking"

The important implication of the evidence above for transport policy is that we can change behaviour by changing the circumstances where people live. If parking in a particular area is reduced *and effectively controlled*, this will reduce car ownership. If this also seems a statement of the obvious, why do some people try to deny it? Apart from the 'evidence we don't want to hear' syndrome, the reasons are partly political, and in the UK, historical.

Until the end of the 20th century, local authorities specified minimum parking requirements for new developments in their local plans. The Labour government of the late 1990s and early 2000s was faced with two incompatible demands:

1. Population, households and the demand for housing were all rising faster than the housing stock.
2. At the same time, opposition to building on greenfield land was growing.

The obvious solution was to intensify, to make more efficient use of development land. Planning guidance was revised, with targets for brownfield development, housing densities and national maximum parking standards for residential and commercial development (offices, retail parks etc). For residential development a national maximum parking guideline of 1.5 spaces per new dwelling was introduced in 2000. Unfortunately, this guidance fudged the issue of parking controls. Authorities were encouraged to reduce the amount of parking to be made available without introducing on-street parking controls, which are often unpopular. It is easier to reduce parking and car ownership in new developments than in existing residential areas, but most planning authorities are reluctant to control parking unless they are forced to, and developers don't want parking restrictions on the streets outside their new houses.

The consequences of this fudge triggered an inevitable backlash. One of the first local authorities to challenge national planning guidance was Kent County Council. Introducing a new local policy in 2009, Highways Manager Bob White explained:

> "'if we reduce parking we can reduce car ownership and reduce car use'. But he said this approach only worked in locations where there were on-street parking restrictions such as double yellow lines and waiting limits. Elsewhere, the result was 'miserable failure' with people parking vehicles on pavements or blocking neighbours' driveways ... "[133]

The Labour government withdrew the 'maximum 1.5 spaces per dwelling' rule in 2006, leaving parking levels to the discretion of local authorities subject to the principle that:

> "Standards should be designed to be used as part of a package of measures to promote sustainable transport choices and the efficient use of land ... "[134]

The incoming Coalition government scrapped that principle and the remaining guidance on parking in commercial developments in 2011. In making these changes, both governments were influenced by local authorities and the development industry. The Coalition government presented the second change with an "end to the war on the motorist" headline.[135] The interests of developers in this respect are not entirely straightforward. On accessible urban sites where land supply is constrained, developers can often increase their profits by reducing parking and building more houses or flats. In other places, allocated parking may be more important to sell a house or flat. Urban developments with little or no parking have successfully sold in many cities, but as I found from 'off the record' interviews with more progressive developers, this is not really about evidence. There is also what psychologists call 'the false consensus effect'.[136] Most of us at some point fall into the trap of believing that others think, feel and act on similar impulses to ourselves. Jobs in property tend to attract status-conscious people who drive a lot for their work and like to own big cars. Whatever the evidence shows, many of them find it hard to believe that a sane, solvent adult would choose to live without a car.

The tension between parking and housing densities is greatest in London. One group of developers, the Berkeley Group, commissioned a study with the aim of lobbying authorities in London for more 'flexible' parking standards.[137] The authors state that "there is no apparent relationship between levels of car ownership and levels of recorded car use". This was based on a study of just 15 new developments, two of which had part of the data missing. Notwithstanding those weaknesses, the report was used to lobby the Greater London Authority and Islington borough council for more permissive parking policies. Islington, like several inner-London boroughs, has a policy facilitating 'car-free' new housing in areas covered by residents' parking schemes: residents of car-free new properties can't have parking permits. Several developers called for this policy to be scrapped, but the council rejected their arguments.[139]

The RAC Foundation supports some serious and useful research, although the Foundation is linked to a motoring organization, which sometimes influences the conclusions of its reports and the way they are presented.[140] Its report on parking policy offers a slightly more nuanced conclusion:

> "there is no direct evidence to suggest that parking constraints are currently a significant influence on car ownership".[141]

'Significant' was not defined in this context. In the Department for Transport survey mentioned on page 53, 7 per cent of adults across England and Wales living without a car cited "lack of a parking space" as a reason.[142] This does not include people who own one car instead of two, as in Poole Quarter. In most parts of the country, residential parking is neither constrained nor controlled, so it is not surprising that parking only affects a minority of car ownership decisions. In discussing the policies of the Labour government in 2005, the report states:

> [Government guidance] "could be interpreted as reflecting the desire of some to deter car ownership in certain locations. **However, there is no evidence to suggest that it will achieve this.**"

In fact, the study offers no evidence one way or the other, apart from references to the fudging problem described above – where parking levels were reduced without effective controls.[143]

Most academic studies of car ownership have looked at parking only in passing (although a new book, *Parking: Issues and policies*, to which I have contributed a chapter, was awaiting publication as I was writing this book[144]). One international study that focused specifically on the relationship between parking and car ownership, using data from New York City, found that parking supply was the most powerful determinant of car ownership – more than other relevant factors such as income and household size.[145]

To take the clearest example in this country, inner-London boroughs have been tightening both parking standards and controls over recent years, in spite of commercial lobbying. This has contributed to a fall in car ownership and a rise in the proportion of car-free households across inner London, from 50.6 per cent in 2001 to 56.7 per cent in 2011.[146] The total population of inner London also grew by 13 per cent over the same period. Imagine what would happen if the policy of parking restraint was replaced with one which encouraged a growing population in a finite space to own, park and drive more cars.

The title of the Berkeley Group study 'Does car ownership affect car use?' is, if taken literally, ludicrous. It invites us to believe that people who don't own cars might drive as much as those who do. Let us suppose, however, that the study, or any other research, had found no statistical relationship between parking, car ownership and car use. What would that prove?

Plausible conclusions might include:
- Parking capacity in the study area was high enough to make no difference to car ownership.
- The formal parking limits were undermined by ineffective controls.

Whether parking constraints *do*, or whether they *could* reduce car ownership are two different questions. Where parking is limited, some people will go to great lengths to keep a car somewhere else, but these are generally a small minority. Evidence for this comes from the various studies of European car free developments[131] and also from student residences. The University of the West of England (UWE), where I work, does not allow students living on the campus to park there. A survey revealed that one in five kept a car somewhere else in Bristol compared to three-quarters of students living off-campus who had a car of their own:[147] UWE is in a suburban location, surrounded by superstores with big car parks, where students can park freely until the store managers get wise to the trick. If, however, you live near the centre of a city in a controlled parking zone and you don't have a parking space, you will generally find it difficult and expensive to keep a car somewhere else: most people won't bother.

This chapter has cited several sources of evidence that support the simple explanations that parking constraints can and sometimes do influence car ownership, and that car ownership exerts a very powerful influence on car use. Parking limitations are not the only way to reduce car ownership, but they are the most direct and probably the most effective. Public transport improvements may have a minor influence: the evidence on this is mixed;[148] proximity and frequency of public transport only seem to influence car ownership in larger cities.[149] It may be possible to reduce car use without affecting car ownership, but the task would be more difficult and the size of any reductions would be smaller.

It may be a blunt instrument, but parking restraint is the most direct and – where properly enforced – a highly effective way of reducing car ownership and use. Whether, where and how car ownership *ought* to be constrained is a question which evidence alone cannot answer: policy makers have to make a judgment informed by evidence, but also based on their values.

Some very real differences of interests and values lie behind the largely spurious debate described in this chapter. According to the British Social Attitudes Survey just over a quarter of the population believe that "people should be able to use their cars as much as they like, even if it causes damage to the environment". [150] If some of those people work for public bodies or lobbying organizations, they have probably learned to conceal their true purpose. If your real purpose is to allow unrestrained car use, then one way to deflect the accusation that you do not care about environmental damage is to promote myths that the most effective policies don't really work.

CHAPTER 7
"You'll never get people over here cycling like the dutch"

Myths:
- Cycling will only ever be a marginal form of transport.
- People only cycle a lot in the Netherlands (and Denmark and Germany – don't mention Switzerland) because it's flat and it's 'part of their culture'.
- You can't transfer ideas from those countries to places with different cultures where hardly anyone cycles – they won't work.
- "Pedestrians and cyclists should generally be accommodated on streets rather than routes segregated from motor traffic."(DfT)[151]

Observations:
- In European cities with the lowest levels of car use, cycling generally accounts for more trips than public transport.
- Rates of cycling in the Netherlands – like most European countries – were falling rapidly until national policy changed in the 1970s.
- Seville, a city with no culture of cycling for transport, followed ideas from northern Europe and increased its rate of cycling tenfold over three years.
- Rates of cycling in Cambridge compare with the best of European practice – and several other UK cities have shown how rapid increases are possible.
- European cities with the highest rates of cycling all have comprehensive networks of separate cycle routes.
- Most non-cyclists in the UK cite separate cycle routes away from traffic as a change most likely to persuade them to give cycling a try.

"Cycling will only ever be a marginal form of transport"

Over three summers from 2006 to 2008, I took off on my bike, cycling 5,000 miles across 7 European countries and visiting cities with low levels of car use, including Groningen, Freiburg, Basel, Copenhagen and Malmö. I was struck by the importance attached to cycling in the transport planning of these cities, the quality and coherence of the cycle routes, and the sheer number of bikes on the streets. I interviewed several transport planners, asking some of them: "How did you manage to achieve all this?" The senior transport planner in Groningen explained that the mayor, councillors and traffic engineers are all regular cyclists, so decisions were made by people who understand. The answer to that question prompted another: "How did cycling become 'normal' there in the first place?" As I would gradually discover, the lazy assumption of British politicians and even some transport professionals that it was 'just part of their culture' was a myth, used to justify another: that things could never work like that over here.

Coming from a country like the UK, it's not easy to appreciate how cycling can be central to transport policy. Figure 7.1 shows the proportion of people who cite cycling as their main means of daily travel in a number of EU countries.

Switzerland was not included in the Figure 7.1 survey: national rates of cycling in Switzerland are slightly lower than Germany but higher than the EU average.[153]

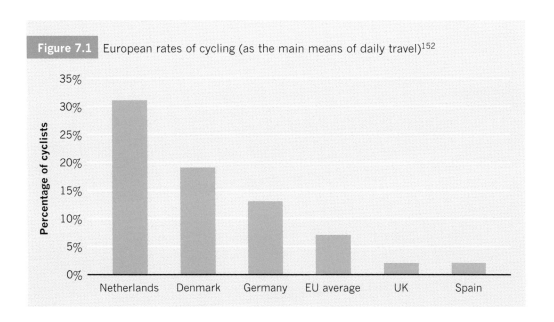

Figure 7.1 European rates of cycling (as the main means of daily travel)[152]

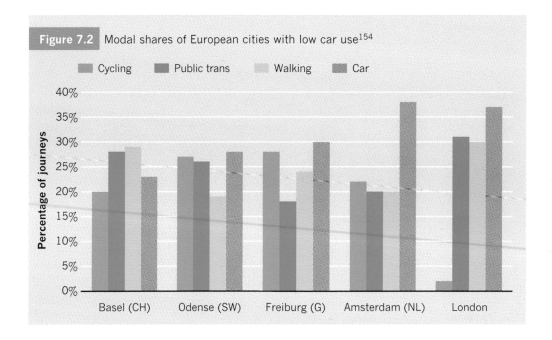

Figure 7.2 Modal shares of European cities with low car use[154]

The greatest potential for cycling as a means of transport is to cover those short-to-medium trips, where rapid travel by other means is not available or not appropriate, particularly in urban areas.

If we look at cities (Figure 7.2) rather than countries, the importance of cycling is clearer. In three of the four continental cities, cycling features in more trips than public transport; in all cases, cycling and walking together account for substantially more.

Comparable statistics for British cities are not usually available: the census only asks about travel to work. One exception is London, where Transport for London (TfL) conducts its own surveys. London has comparably low car use, but mainly due to high use of public transport, as discussed in Chapter 14. The low level of cycling partly reflects UK transport policy over the years, but cycling and public transport may also substitute for each other. The public transport network in London is more comprehensive, catering for a much greater range of travel needs than would be possible in a smaller city. Although the rate of cycling in London is very low by European standards, increasing it has become an important objective for TfL, particularly in the central areas, where the tube network often reaches capacity, with some station barriers closing to keep passengers out at peak times. The upgrade to the underground network that began in 2010 was budgeted to cost £30bn – nearly £5,000 for every adult living in the capital. If many more people would cycle instead of using the tube at peak times, it could save a lot of money.

"People in Europe cycle because it's flatter, warmer and dryer"

Three of the European cities above are relatively flat. Freiburg is mainly flat, although some of its suburbs creep up the foothills of the Black Forest. A flat terrain undoubtedly encourages cycling, but it is not a determining factor. There are hilly cities with many cyclists and flat cities with few cyclists. Heidelberg, which has mountains within its boundaries, has a 25 per cent share of cycling. Bristol, which is hilly, has recently increased its share of cycling to work to 8 per cent (high by British standards), whereas Coventry and Northampton, which are both much flatter, only achieve 2.8 per cent.[155]

Another favourite myth, or excuse for inaction, relates to weather: of course, people in all these other places can cycle because they have better weather than us. In reality, Amsterdam has similar rainfall and average temperatures to Manchester; Danish cities are considerably colder.[156] Across Europe, it is the colder, wetter countries of the northwest where people cycle more than the warmer, dryer countries around the Mediterranean.

So why have some cities and countries been so much more successful in encouraging cycling than others? As with most transport problems, part of the answer is very simple, and part is rather complicated. The first and most obvious reason relates to cycling infrastructure. Study after study in the UK has found the main factor deterring non-cyclists is the fear or dislike of mixing with traffic. The main factor that would persuade them to give it a try would be continuous separate cycle routes, which protect them from traffic.[157] A few who have tried cycling in one of those European cycling cities may be speaking from experience. As one young woman I interviewed put it:

> "When I was living in Germany, I cycled in Bonn and that was absolutely blissful because there were enormous wide cycle lanes along the side of every road, sectioned off from the traffic by the little mini islands, or whatever – bollards in the middle of the road, or whatever, and they were wide and they were comfortable and they were smooth and there were no potholes and nobody tried to kill you ... "[158]

On returning to London, where conditions were rather less blissful, she gave up cycling.

'Filtered permeability'

After returning from my trips around Europe I wrote several articles for cycling magazines and *Local Transport Today* challenging the conventional thinking on cycling policy.[159] One of these articles coined a term that would gradually enter the lexicon of transport planning: 'filtered permeability'.[160]

The separation can come in many forms such as cycle paths, bus gates or footbridges. The advantage is usually in the form of a short-cut, though it may also save time or increase comfort – by avoiding a congested stretch of road or a steep hill, for example. This principle, which I observed across most of the European cycling cities, was very different to any which guided the UK approach (which is based on shared streets, shared pavements and a presumption against separate cycle paths). The few voices raised against the British approach were marginalized in the mid-2000s: it would be a few years before they were taken seriously in mainstream UK transport debate.

Before my 2007 trip around the Netherlands I used to sympathize with the argument that 'you can't build cycle routes everywhere'. Separating cyclists from traffic does not mean building separate cycle paths on every street: separation can be achieved by strategically blocking streets to through traffic. The Dutch (in particular) have gradually built a network using a combination of quiet streets and separate cycle paths, which enables most people to avoid traffic almost all of the time – and to do so without significant detours. Indeed, cycle routes are often more direct than the alternatives for general traffic. Transport authorities in some of the larger Dutch cities measure the proportion of journeys where the bike is faster than the car, with the aim of increasing this over time.

This is the principle of filtered permeability. Cities like Groningen and Freiburg have built a fine-grained network of cycle routes and at the same time have limited the number of through routes for motor traffic. Closing their city centres to through traffic with extensive pedestrianization has helped this process. Natural barriers such as rivers and manmade barriers such as railway lines are crossed by bridges and underpasses for cyclists and pedestrians only. Roads are blocked off to through traffic but kept open for cyclists and pedestrians in a coherent way, creating on-road cycle routes with long straight stretches (see Figure 7.3) instead of the continual twisting and diverting typical of British cycle routes.

Cycling infrastructure is never perfect, in any city, but there is greater consistency in design standards across northwest Europe: wider paths, higher design speeds and

Filtered permeability
Separating different modes of transport to give an advantage to some modes over others.

Figure 7.3 Filtered permeability in Malmö provides a direct route to the city centre

greater priority over side roads and general traffic. The shared pavement, stopping at every side road or driveway, can still be found in some parts of Germany but has been gradually disappearing across the Netherlands and Denmark. Three-way separation of cyclists, pedestrians and general traffic is a key principle in most of these cities, with on-road cycle lanes being progressively replaced by hybrid paths (Figure 7.4).

Consistency of design standards is critical to the success of certain approaches such as the cycling priority roundabout commonly found in Dutch and Danish cities (Figure 7.5). Some attempts have been made to introduce similar designs in Britain, but problems occur when drivers are not familiar with designs that give priority to cyclists. A half-hearted copy of this design was attempted on a road section of the National Cycle Network in Ivybridge, Devon a few years ago. It was built then swiftly removed after a series of 'near-misses' involving drivers who were not expecting, or did not understand, that cyclists on the outside had priority over them.

The health and safety culture in British authorities has contributed to this short-term, shortsighted decision-making. Cycling officers complain that attempts to provide priority over side roads are sometimes rejected on the advice of safety auditors, whose sole concern is liability for the scheme they are auditing at the time, ignoring the broader, longer-term implications of their decisions.

66 URBAN TRANSPORT without the hot air

Figure 7.4 Hybrid path in Katwijk aan Zee, Netherlands

Figure 7.5 Roundabout with cycling priority, Esbjerg, Denmark

Another reason people in the UK often cite as a reason for not cycling relates to sweating while cycling and the absence of showers at work. This often surfaces in discussion among travel planners and in focus groups or workplace surveys.[161] Unfortunately, employers who install showers often find they make little difference. Showers are mainly used by existing cyclists in a culture where Lycra, rucksacks and competitive riding are all more common among the small group of people – mainly men – who currently cycle to work. In the cities and countries of northern Europe, sedate cycling in normal clothes is more common. Helmets are rare and casualty rates are lower in countries that have found safety in numbers. Cycling is normal for both sexes there, whereas the male dominance of cycling in Britain reflects and helps to perpetuate the harsher street environment.

Another favoured excuse among British highway engineers for inaction or poor quality facilities is that "continental cities have wider main roads". This is true in some places, although much of that width has been taken away from general traffic (Figure 7.7). Opportunities in the UK are often overlooked or ruled out if they might reduce capacity for general traffic (Figure 7.6).

Where sufficient width is not available, particularly in older historic areas of cities like Amsterdam, selective road closures or one-way streets with cycling contraflows are widely used. Cycle paths through pedestrian areas are more common (Figure 7.8), allowing cyclists to continue at their normal speeds with fewer interruptions – anathema to many British transport professionals, who prefer sharing of space, with all the conflicts, delays and discouragement to cycling (and walking) which that entails.

Figure 7.6 "British roads aren't wide enough to build proper cycle paths": Bristol's inner ring road

Figure 7.7 Groningen's inner ring road

Figure 7.8 Cycle path through pedestrianized area in Odense, Denmark

Different ways have been found to give priority to cyclists at junctions, like the cyclist-only traffic light phasing in some Dutch cities: all other traffic stops at one point in the cycle, while cyclists are allowed to move in all directions. As consistent priority is given to bikes, jumping of red lights is not as frequent as it is in British cities, where it has become normal behaviour.

Cycle parking is taken seriously in European cycling cities, although there isn't always enough of it. Elaborate underground or multistory staffed cycle parking facilities are available in the centres of most European cycling cities – often by the main railway station (Figure 7.9 and Figure 7.10).

"Ideas from overseas won't work where hardly anyone cycles"

This brings us back to the issue where this chapter began. How did cycling become normalized in several European countries and why not in the UK? In the early part of the 20th century cycling was a common means of transport almost everywhere

CHAPTER 7 You'll never get people over here cycling like the Dutch

Figure 7.9 Underground cycle parking beneath Basel Railway Station

Figure 7.10 Cycle parking by Amsterdam Railway Station

in Europe. With rising car ownership, rates of cycling began to fall everywhere, but most rapidly in the country where its influence on national culture was strongest – the Netherlands. Then in the 1970s, two factors combined to force a change in Dutch transport policy:
1. The first oil crisis of 1973-4.
2. Growing public concern at the rising death rates on the roads. In 1973, a campaign called 'Stop the Child Murder' was launched, which achieved a high media profile and had a considerable influence on politicians.

This change is sometimes presented as an abrupt one, but the effect on official thinking and national policy was actually more gradual. The 1989 Dutch National Transport Strategy was more explicitly pro-cycling than the 1980 version. Although progress was variable, transport policy across the country shifted towards greater separation of cyclists, and restraint of motor traffic in urban areas. The two are seen as corollaries, not alternatives as they are sometimes presented in the UK.

The results can be seen in Figure 7.11: cycle use reached a low point, and cycle fatalities a high point, in the 1970s; after that, both continued to move in the right direction for the next three decades.

Danish politicians followed the Dutch lead with high-level Danish delegations to the Netherlands during the 1980s. Similar principles were followed in a number of Danish and German cities. Spectacular increases in cycling were seen in some cities such as Copenhagen, Odense and Freiburg, although national rates of cycling remained stable in Denmark and Germany; increases in the bigger cities were balanced by gradual decline elsewhere.[161i]

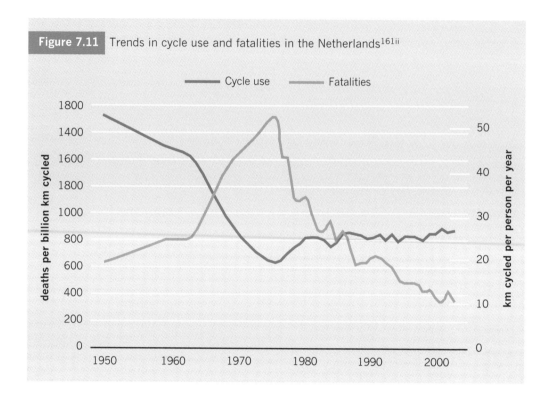

Figure 7.11 Trends in cycle use and fatalities in the Netherlands[161ii]

The argument that 'it's just part of their culture' makes a common mistake in social analysis: confusing absolute differences with reasons for change. Cycling has always been more prevalent in the Netherlands, helped by its flatter terrain, but this does not explain why rates of cycling recovered there, while they continued to fall in the UK. Most analysts who have looked at the history and practice of successful European countries have come to similar conclusions about the factors that encourage cycling: better infrastructure, traffic restraint and a consensus between politicians and transport professionals to make cycling a priority.[161iii]

But as all of these changes happened alongside the growth in cycling in the Netherlands and Denmark, we are back to the 'what causes what' problem. And even if the changes did cause an increase in cycling, would they work somewhere else? Another example from a totally different part of Europe can help answer both of these questions. This is the story of Seville, a city with no cycling culture until the early 2000s, in a country where cycling was mainly seen as a sport rather than a means of transport. In 2003, a change of political leadership brought to power a coalition determined to change the city's transport culture.

Through a combination of a Dutch-style network with segregated cycle routes (also influenced by the principle of routes for ages 8 to 80 from Bogotá in Columbia), removal of traffic from the city centre, and a city cycle hire scheme copied from several French cities, Seville increased the proportion of journeys by bike from 0.2 per cent to 6.6 per cent by 2009.[161iv] As in London, these journeys are particularly concentrated on routes in and out of the city centre, so the effect on the environment of the city is much greater than those figures suggest. My brother lived in Seville in the 1990s, when only madmen and foreigners would cycle for transport around the city; he found a radical change in attitudes towards cycling when he returned there after 2008. The above changes were implemented at the same time as a new tram network, and some writers have pointed out that the cycle routes carry more trips at a fraction of the cost. That may be true, but as we shall see with other examples in Part II, integrated changes reinforce each other: it was easier for the politicians in Seville to justify the removal of road space and the changes to the city centre as part of an integrated package. Note that the growth in cycling and cycling culture followed all of these changes, so although we cannot measure the effect of each element, the package clearly *caused* the growth in cycling – not the other way round.

These changes were no easier to implement in Seville than they would be in British cities. A significant minority vociferously opposed them. Shopkeepers and some local residents protested over the loss or relocation of parking spaces. Some people even filmed themselves vandalizing the hire bikes and then uploaded the evidence to You Tube. Two studies reveal an interesting parallel between the political attitudes of the people who are hostile to cycling in Seville and in Britain. In Seville, the people who most opposed the cycling changes also favoured retaining the street names that commemorate leaders of the Franco dictatorship.[161v] Although the pattern is not as simple as a leftwing/rightwing distinction, people who are most hostile to cycling in Britain tend to vote for the (anti-European) UK Independence party (UKIP).[161vi] This may explain the anti-cycling stance of the candidate cited in Chapter 3, page 25 and the similarities in the treatment of cyclists, immigrants and foreigners in newspapers such as the *Daily Mail* (discussed in Chapter 3), which is widely read by people who vote for that party.[161vii]

"Shared streets are better for cycling than separate paths"

To understand how the UK took the wrong route on cycling, we again need to look back into history. Like many of our national failings, this one was partly rooted in the English class system. Before the Second World War, when cycling was much more common than it is today, the Cyclists Touring Club (CTC) was the most important national cycling organization. It was dominated by people for whom cycling was a passion and a leisure pursuit. They yearned for the freedom of the open road, viewed cycle paths as a conspiracy and paid little attention to the growing cohort of urban utility cyclists drawn mainly from the lower classes. In a letter to *The Times* in 1934, their national secretary wrote:

> "**The demand for separate tracks for cyclists is part of the campaign of motorists to appropriate public highways for their exclusive use.**"[161viii]

In other words, they feared that cycle paths would lead to the banning of cyclists from major roads (which has occurred in several European countries). The CTC lobbied against the separation of cyclists from general traffic and largely achieved its aims. At a time when cities were expanding and many new and wider roads being built, very few included any separate provision for cyclists.

Rates of cycling fell rapidly after the Second World War as car ownership grew across most of the developed world including Britain. In the late 1970s, when several countries were re-evaluating their transport policies, a small group of volunteers in Bristol started work on a project which would have wider national implications. The Bristol to Bath cycle path was created by Sustrans (an NGO which promotes sustainable transport) in 1979 and became the first of many such projects around the country. It remains one of the best examples of its type, providing direct, almost uninterrupted cycle access towards the centre of both cities, equally useful for utility and leisure cycling (Figure 7.12).

The highest transport priority of the Thatcher government in the 1980s was road building. Cycling was viewed as an irrelevance. The speeches of Margaret Thatcher (searchable on the website of the Margaret Thatcher Foundation[16ix]) contain 487 references to roads or cars and only one to cycling. But by the mid-1990s, under John Major's Conservative government, attitudes were changing. In 1995, Sustrans submitted its application to create a national cycle network to the Millennium Commission and the following year the Department for Transport(DfT) published the National Cycling Strategy. Written in consultation with cycling organizations, this set a target to quadruple cycle use by 2012, mainly at the expense of travel by car. "Cycling has a bright future" it concluded, optimistically.

The Strategy included many useful recommendations but suffered from several flaws which would also afflict the National Cycle Network. By its own standards, the Strategy was an unmitigated failure, as illustrated in Figure 7.13.

The Strategy was national but only advisory: implementation decisions on cycling policy and infrastructure have always been taken by local authorities and at this level the transport culture and priorities were changing slowly, if at all. Cycling and walking are the two least expensive forms of transport, but the proportion of the

CHAPTER 7 You'll never get people over here cycling like the Dutchg

Figure 7.12 Bristol to Bath cycle path

transport budget allocated to either remained much lower than in the countries of northern Europe. Precise measures are difficult to obtain, but spending on cycling never exceeded 1 per cent of the infrastructure budget.[162] National governments launched various one-off funding programmes, with great publicity, over the years, but core budgets continued to marginalize cycling and walking. Whereas most of the national transport budget, particularly on roads and rail, is spent nationally, the vast majority of public spending on cycling is paid from central government to local authorities. This can create a false impression at the local level that the share of transport spending allocated to cycling is significant or even excessive according to some opponents.[163]

Cyclists who had survived the excesses of Thatcher's 'great car economy'[164] were a hardy and unrepresentative bunch – overwhelmingly male. Hostility to segregation remained strong among many cycling organizations and activists.[165] In 1996, several of these organizations collaborated with the DfT and the Chartered Institute of Highways to produce semi-official guidance on the design of cycling infrastructure, which included a hierarchy of provision favouring cycling on streets shared with general traffic.[166]

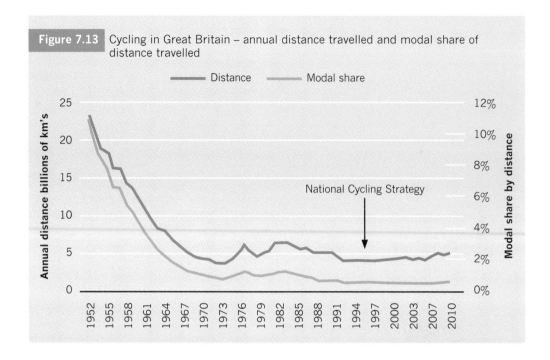

Figure 7.13 Cycling in Great Britain – annual distance travelled and modal share of distance travelled

The National Cycling Strategy fell quietly into disuse and national targets for cycling were formally abandoned in 2005, but the hierarchy which had contributed to its failure lived on. It was adopted in similar forms in successive government guidance documents including the 2012 version shown in Figure 7.14, which remains UK national policy.

This hierarchy ignores the success of experiments like the Bristol to Bath railway path and prefers paint on the road to separate, continuous off-road routes. The

Figure 7.14 Hierarchy of provision for cyclists in British national guidance[167]	
Consider	Possible actions
first ↓ last	Provide for cyclists in the carriageway — Traffic speed / volume reduction; HGV reduction; Junction / hazard site treatment; Reallocation of carriageway space
	Create new shared use routes
	Convert pedestrian routes to shared use

phrase 'shared use' means shared with pedestrians. The possibility of separate paths for cycling (like the hybrid path in Figure 7.4) is not even mentioned. The lowest priority given to shared footways makes more sense but has not prevented their proliferation. The shared-space and New Urbanist movements, discussed in the next chapter added further obstacles with the publication by the DfT of *Manual for Streets* in 2007.[168] This reiterated the preference for shared streets but added that off-road paths were acceptable if they were short. This combination of shared streets and short paths has reinforced the culture of ad hoc compromise in the design of cycle routes in the UK.

A national guidance note on design for cycling contained another flawed compromise which continues to constrain cycling in Britain – the recommendation that authorities should plan for different types of cyclist, as follows:

> "Some cyclists are more able and willing to mix with motor traffic than others. In order to accommodate the sometimes conflicting needs of various user types and functions, it may be necessary to combine measures or to create dual networks offering different levels of provision, with one network offering greater segregation from motor traffic at the expense of directness and/or priority."[169]

Authorities following this approach typically provide two types of infrastructure: haphazard strips of paint on the road for serious cyclists and shared pavements conflicting with pedestrians for everyone else (Figure 7.15). Cycle routes in Britain are typically longer and have more frequent changes of direction than the alternative routes by road.

Research conducted among non-cyclists and occasional cyclists reveals that two key factors affecting their attitude to cycling are the dangers of mixing with traffic and time pressures combined with a perception that the car will usually be quicker.[170] When asked what would help reduce their fear of traffic, the top answer in many surveys is separate traffic-free cycle routes. This is not just about accident statistics – *being* safe is not enough: to attract new cyclists, routes must also *feel* safe and comfortable to ride on. This is particularly important for women, who cycle less than half as often as men in Britain.[171] In European cycling cities, there is more of a gender balance.

A difference in the legal liability for road casualties is often cited as another explanation for higher rates of cycling in the Netherlands and Denmark, particularly by people who want to downplay the importance of segregated infrastructure. In fact, the importance of this difference has been exaggerated: the burden of proof in civil cases where cyclists and pedestrians are injured by motor vehicles is lower in the

Netherlands (and a similar move could encourage more responsible driving in Britain) but this change was introduced in the 1990s, many years after the Dutch cycling renaissance began.[172] It was a consequence of that renaissance, not a cause.

In providing segregated routes in the UK, the design guidance advises transport planners to sacrifice the desire to save time, which is a key reason why occasional cyclists usually choose to drive. Although they may not realize it, design speeds and continuity may be more important for slower, less confident cyclists, than they are for those who are fitter and more reckless. In European cycling cities with traffic-free routes that are more direct than roads for general traffic (eg Figure 7.16), all types of cyclist use them.

The poor quality of cycle infrastructure in the UK has produced a vicious circle: many cycling organizations have supported this flawed compromise because they feared any alternative might threaten their right to ride on the road. Whereas many European countries including the Netherlands ban cycling on a wider range of main roads than the UK, defending the dubious 'right' to cycle on these roads has allowed UK authorities to avoid providing effective alternatives.

I might have finished this chapter on that fairly depressing note were it not for a few small but significant signs of positive change. Although cycling nationally was plateauing, a number of British cities were showing some impressive rates of increase. My home city of Bristol increased cycle commuting by two-thirds between the 2001 and 2011 censuses.[173] Cambridge, which has some of the best cycling infrastructure by British standards, produced figures to compare with the best of European cities: a third of commuters cited cycling as their normal mode, and in another survey, just under half of adults cycled at least once a week.[174] Following many visits from curious foreigners, the Dutch Cycling Embassy was set up in 2011 to disseminate best

Figure 7.15 UK: two types of substandard provision

Figure 7.16 Denmark: quicker route for all types of cyclist

practice to other countries. This was followed by the Cycling Embassy of Denmark and the Cycling Embassy of Great Britain, an organization which campaigns for "dedicated safe cycle infrastructure, in line with the best practice found around the world".[175] The CTC, which had strongly supported the hierarchy of provision (Figure 7.14 with minor variations) from 1996 onwards, published a new policy, more favourable towards continental-style separate facilities, after consulting its members in 2012.[176] As its campaigns and policy director explained, many CTC members have cycled in other European countries and have come home wondering why we can't build cycle routes like that in Britain.

In central London, the congestion charge and terrorist attacks on London's public transport network gave a significant boost to cycling from 2003 onwards. The significance of this was exaggerated in some media commentary: across London as a whole, rates of cycling remained stubbornly low, but the concentration of cyclists in parts of inner London and around the seat of government in Westminster raised the issue up the agenda of politicians and the national media. Following a serious injury to a reporter cycling to work, in February 2012 *The Times* launched its influential 'Cities Fit for Cycling' manifesto and has given prominence to cycling issues ever since. In the same month, the London Cycling Campaign launched its 'Love London, Go Dutch' campaign in the run-up to the London mayoral elections in May 2012. Many people were sceptical when Mayor Boris Johnson endorsed the campaign, until shortly after his re-election he produced a rather surprising policy document (*The Mayor's Vision for Cycling*) committing £913m to cycling over 10 years, comparable to levels of spending in the European cycling cities. Although future spending may be difficult to guarantee (and there will be two mayoral elections in those 10 years), the most significant departure from the past can be found in the language of this document. Admitting that existing cycle infrastructure left much to be desired, it made the bold statement that:

> "Timid, half-hearted improvements are out – we will do things at least adequately, or not at all."[177]

This policy went directly against the national guidance by declaring that "we will segregate where possible". This would include Dutch-style hybrid paths and a 15-mile 'Crossrail for the bike' – a continuous cycle route crossing London from east to west.

Time will tell to what extent the reality will match up to these commitments, but from 2012 onwards there were already signs of a shift in attitudes among transport planners and some political leaders in other British cities. These new attitudes brought talk of 'Dutch-style' cycling infrastructure,[178] which various government programmes would help to fund; but as the first examples began to appear in 2014 the reality was, unfortunately, failing to match the hype.

Figure 7.17 A 'Dutch-style' cycle path under construction in Bristol

One of the new generation of 'Dutch-style' cycle paths under construction in Bristol (Figure 7.17) illustrates the problem. If it really were Dutch it would continue in a straight line with priority over the side road in the background, instead of which, it diverts left on to a crossing shared with pedestrians. This new generation of cycle routes was meant to separate cyclists and pedestrians, but designs of this type force the two together at junctions, where potential for conflict is greatest. They also delay cyclists and interrupt their journeys, causing many of them to stay on the road instead.

The reason for this failure to match continental good practice relates to the culture of the safety audit mentioned earlier. Traditional British highway engineers claim that motorists would fail to stop at junctions where cycle paths have priority despite the evidence from the Netherlands and Denmark, which both have much lower rates of cycling casualties than the UK.[179] (The excuse that they can do it 'over there' because they have different laws is, as we saw earlier, a myth.[180]) How local politicians and transport planners might overcome some of these problems and get serious about cycling in urban areas will be discussed in Chapter 16.

CHAPTER 8

"The car can be a guest in our streets"

Myths:
- Shared space:
  - Reduces traffic speed.
  - Reduces pedestrian casualties.
  - Encourages walking and cycling.
  - Can solve the problem of pedestrianized areas "becoming lifeless places at night".[181]
- 'Permeable' street networks, which allow cars to move in all directions, "encourage walking and cycling".[182]

Observations:
- The evidence on shared space, vehicular speeds and pedestrian casualties is ambiguous. Several of the claims about shared space reducing speeds and casualties are based on misinterpretations of evidence.
- There is no evidence – and no reason to believe – that shared space encourages walking or cycling.
- In some circumstances shared space can create a hostile environment for pedestrians, particularly older people, women and people with disabilities.
- Traffic cannot create life on streets; only people can do that. Lifeless pedestrian areas are caused by retail-only planning restrictions, which can be changed.
- Permeability has little impact on travel behaviour unless it is filtered to favour some modes over others.

"Shared space slow cars, reduces casualties and encourages walking and cycling"

If you study or work in transport planning or urban design, you will be well aware of the shared-space movement and the impact it has had on conventional thinking about street design, particularly in the UK. If not, you may have heard the term 'shared space' describing a street in your town, or a few high-profile examples like Exhibition Road in west London (Figure 8.1).

The DfT defines shared space as follows:

> "A street or place designed to improve pedestrian movement and comfort by reducing the dominance of motor vehicles and enabling all users to share the space rather than follow the clearly defined rules implied by more conventional designs."[183]

If you had never seen a shared-space street (or Figure 8.1) would that definition give you any idea of the types of design the DfT was referring to?

A street "designed to improve pedestrian movement and comfort" sounds a pretty good idea. We have many examples: they are called pedestrianized streets. But

Figure 8.1 Exhibition Road, west London: an example of shared space

shared space implies something completely different. Whereas pedestrianization improves the comfort of pedestrians by removing motor traffic, shared space aims to mix the two: the more they mix, the more 'successful' the shared space. The only clue as to what the writers actually mean is the phrase "clearly defined rules". Under shared space, it implies, rules are not clearly defined. A simpler and clearer definition of shared space is:

> **Shared space**
>
> "streets designed to minimise demarcations between vehicles and pedestrians" (Moody and Melia 2013[184])

The types of demarcation that shared space seeks to remove typically include: kerbs, pavements, pedestrian guardrails, traffic lights and other traffic signs. In its purest forms, like Elwick Square on the inner ring road of Ashford in Kent (Figure 8.2), the street between the buildings is a flat featureless surface. The concept is best understood as a design approach rather than a black and white definition: a shared-space conversion is one where at least some demarcations have been removed.

Figure 8.2   Elwick Square, Ashford, Kent (Photo: Simon Moody)

Figure 8.3 Laweiplein 'Squareabout', Drachten, Netherlands

I met the originator of the shared-space concept, Hans Monderman, in the town of Drachten in the Netherlands in 2007, shortly before his untimely death. At the time, Monderman was little known in his own country. A transport planner I met in Amsterdam told me: "I'd never heard of this man until foreign visitors started mentioning him." As the district's chief traffic engineer, Monderman had two problems he wanted to solve at certain junctions: high rates of casualties, and traffic congestion. The first, high-profile, scheme replaced traffic lights with a quasi-roundabout nicknamed the 'Squareabout' at the Laweiplein in central Drachten (Figure 8.3).

This succeeded in easing the congestion, and the casualty figures fell over the following years (although a later study questioned whether the sharing of space was actually the cause of this).[185] The idea was applied in a slightly different way to a few other junctions and to some quiet residential areas in a way that's familiar to residents of new estates in many parts of Britain where developers prefer not to pay for pavements.

I noticed that like most Dutch towns, Drachten has a comprehensive network of segregated cycle routes (Figure 8.4). Monderman's colleagues were very clear – the two concepts went together: shared space was not intended as an alternative to segregated cycle routes (as it is sometimes presented in Britain).

CHAPTER 8 The car can be a guest in our streets

Figure 8.4 Segregated cycling infrastructure, Drachten, Netherlands

Monderman had no idea whether the few shared-space schemes around Drachten had increased or reduced walking or cycling; that was never his intention and he had no reason to expect that they had. Only after crossing the North Sea did shared space acquire the dubious status of a sustainable transport measure, a tool for modal shift from driving towards active travel.[186]

MVA Consultancy surveyed practice around Britain (research commissioned by the DfT) and found that shared-space schemes were implemented for the following purposes:

- improving the urban environment;
- giving people freedom of movement rather than instruction and control;
- improving the ambience of places;
- enhancing social capital;
- enhancing the economic vitality of places;

to which I sometimes add, when lecturing on this subject:

- The achievement of the kingdom of heaven on Earth!

So why did a prosaic and technical approach to street design inspire such fervent beliefs? Is there any evidence to support any of these claims?

A clue to the first question lies in the DfT's eulogizing definition with its reference to "clearly defined rules". Some advocates of shared space have presented it as a libertarian measure, freeing motorists from the tyranny of controls like traffic lights.[187] Others have argued that removing formal controls will induce motorists and other road users to behave in a more civilized way towards each other.[188] In an era dominated by the neoliberal ideology of free markets and deregulation, these arguments have proved very persuasive. Following sympathetic national guidance, local authorities around the country have pursued shared-space schemes with varying degrees of enthusiasm, frequently overriding the concerns of more sceptical residents and organizations representing disabled people.

In the general enthusiasm for the idea, many transport professionals have overlooked some fairly obvious shortcomings in the evidence used to support it. The claim that shared space reduces casualties depends on what is known as 'risk compensation theory'. Environments that feel safe encourage risk-taking, whereas environments that feel risky make people more cautious. There is some evidence of this effect among road users. But if sharing space makes motorists feel more at risk, we would also expect a similar effect on pedestrians and cyclists. If we design streets to make pedestrians and cyclists feel less comfortable, we might make them take more care, but we might also inadvertently discourage them from walking or cycling.

Of all the claims made about shared space, the one that it causes drivers to walk or cycle instead[189] is the easiest to dismiss. There is no evidence that it reduces car use or increases active travel, nor is there any reason to believe that it might do so, unless it is accompanied by other measures such as reducing road capacity or speed limits.

The claims about casualty reductions are more difficult to assess. Shared-space schemes are often accompanied by other measures such as reduced speed limits, narrowed carriageways or redesigned junctions. Reductions in casualties may be due to these other factors rather than the removal of demarcations to create shared space. A clear example of this occurred on the inner ring road of Ashford, Kent, where I supervised a research project conducted by Simon Moody, a transport planner studying at UWE. Both the speed limit and the width of traffic lanes were reduced, roughly halving the volume of traffic. Over the following three years casualties fell by 41 per cent.[190] Whether removing demarcations contributed to that improvement or hindered it is impossible to say, but a simple headline can often carry more weight than a nuanced report. The Ashford example has been widely cited as a success story, and at least one other local authority has used it to support its own shared-space scheme, where public opposition was partly motivated by safety concerns.[191]

Surveying the literature on this subject, MVA Consultancy, which was advising the DfT on its national guidance, concluded that the evidence was mixed, a message which could have proved politically awkward. Spot the difference between these two statements:

> "There is no evidence that Shared Space schemes ... as implemented in the UK have more casualties ... There is some evidence from the Netherlands that, at locations with greater than c.14,000 vehicles per day, Shared Space layouts may have more casualties, relative to traditional layout ... "
>
> MVA (2009)[192]

> "Available evidence indicates a comparable number of casualties in shared space streets and conventional streets ... "
>
> DfT (2011)[193]

MVA was also commissioned to do its own research to inform the guidance which claimed to be "evidence based".[194] This produced some plausible findings and some which were rather suspect. The claim that sharing space causes drivers to slow down was based on a survey of 10 sites around the UK. The ones with fewer demarcations tended to have lower traffic speeds. The report's authors concluded that fewer demarcations *caused* lower speed but the opposite explanation was just as plausible. If you were a highway engineer, would you recommend removing signs, kerbs, pelican crossings etc on a road where speeding was a serious problem? If not, then fewer demarcations might be an *effect* of lower speeds rather than a cause – or possibly some element of both depending on local circumstances. Despite this rather obvious problem, the national guidance makes the following statement:

> "As the degree of 'sharedness' (ie the physical aspects of a street that encourage sharing) increases, vehicle speeds tend to reduce."[195]

Shared spaces work better for pedestrians where traffic volumes and speeds are low. MVA's finding in that respect is corroborated by a range of other studies. More pedestrians on the street may also help. This is the case in places like New Road in Brighton (Figure 8.5), which feels almost like a pedestrianized street, particularly during the daytime when there are far more pedestrians than vehicles. Under these circumstances, vehicles *are* forced to slow down, but the decision to treat New Road in this way was partly influenced by low traffic volumes in the first place.

By contrast, in Elwick Square in Ashford, vehicles substantially outnumber pedestrians. Removing all demarcations there produced an environment which most pedestrians found hostile. Of those surveyed, 80 per cent said they felt safer under the previous street layout – despite the higher traffic volumes. Video observations

Figure 8.5 New Road, Brighton, a semi-pedestrianized shared space

showed people giving way to vehicles far more often than vehicles giving way to people. Many people were seen running across the square. A few said they knew people who avoided the square altogether. Like several other studies of shared spaces, we found young men to be more positive, while older people and women, particularly with young children, disliked the forced interaction with moving vehicles.

Although wheelchair users tend to prefer flat surfaces, several studies have reported negative experiences among people with a range of disabilities.[196] Guide Dogs for the Blind has consistently opposed shared-space schemes, based on some research of its own and feedback from users of its services.[197] The problems suffered by blind people in shared-space streets were graphically illustrated by the documentary *Sea of Change*, screened at events around the UK.[198]

Some of the strongest criticisms in Ashford were reserved for the 'courtesy crossings' – faded zebra crossings built into the road surface, without the orange warning lights of normal pedestrian crossings ('Belisha beacons'[199]). To cross the road most people did use them, rather than risk crossing the centre of the square, but only a minority of drivers gave way to pedestrians seeking to cross there. These problems have not prevented other authorities from installing similar crossings on shared-space streets, provoking some public criticism.[200]

Several studies have found that, when offered the choice, pedestrians prefer pedestrianization to shared space.[201] One of the reasons I became concerned about the growing enthusiasm for shared space in Britain is that it has been promoted as an alternative to pedestrianization in several places. There is no evidence that shared space has achieved any positive modal shift but there is a long-standing, solid body of evidence that pedestrianization encourages people to walk[202] (it can also encourage cycling providing cycling access is maintained, ideally through separate paths as illustrated in the previous chapter).

"Shared space is a better alternative than lifeless pedestrianized areas"

The word 'pedestrianization' often evokes images in Britain of paved concrete shopping areas built in the 1960s and 1970s, influenced by brutalist architecture. Their ugliness has tarnished the concept but, as we shall see in Chapters 12 to 15, pedestrianization has also enhanced some of the most beautiful urban spaces in British and European cities.

According to the DfT, shared space:

> "also addresses a particular problem which can affect some pedestrianization schemes, where the absence of vehicular traffic can lead to them becoming lifeless places at night". [203]

The problem of "lifeless places at night" is not caused by pedestrianization, nor can it be solved by reintroducing traffic: it is caused by planning policies which designate pedestrianized areas for retail use only. Many councils still act on the mistaken belief that people cannot (or should not) live in places without parking or direct vehicular access, even though there are several successful examples of residential development within pedestrianized areas (discussed in Chapter 13 on carfree developments). The other obvious solution is to introduce more venues which are open at night such as restaurants (Figure 8.6) – a step which some councils seem surprisingly reluctant to take.

Pedestrianization also increases spending and property values on shopping streets.[204] In spite of this, organizations like the Prince's Foundation have advised developers or local authorities to choose shared space over pedestrianization. In at least one project in which they were involved, in Nelson, Lancashire, traffic was reintroduced into a formerly pedestrianized street in the belief that this would boost trade.[205] It would be impossible to *measure* whether reintroducing traffic did boost trade for the same reasons that have misled people over shared space: several changes were made at the same time, including improvements to the fabric of the street – costing a total

Figure 8.6 | Quakers Friars, Bristol: pedestrianized area with restaurants and residential accommodation

of £2.3m.[206] We will never know what might have been achieved if that money had instead been spent on improving the pedestrianized area. By ruling out other alternatives, myths and fashions can create their own spurious evidence.

Some PhD research at UWE (not yet published) provides another perspective on pedestrianized areas. Through a mixture of 'walkalong' and static interviews the researcher revealed ways that people use everyday walking in urban areas to relax, de-stress, solve problems and come up with creative ideas. Most of the participants were also car drivers but even those most sympathetic to urban driving enjoyed the relief of entering a pedestrianized area on foot. Where walkers are uninterrupted, they are able to lose themselves in their thoughts (green areas were particularly useful in that respect). Frequent interruptions, close proximity to traffic, and places where pedestrians must continually watch out for vehicles (which would include shared-space streets) can all break the spell and extinguish the flow of creative thoughts. 'Vehicles' in this context can also include bicycles. Other studies have found considerable resistance among pedestrians to sharing pavements with bicycles[207] – an important finding for the discussion on cycling in Chapter 16.

"Permeable street networks that allow cars everywhere encourage walking and cycling"

The preference for shared space over pedestrianization of the Prince's Foundation and the DfT is linked to another preference for 'streets which are permeable or connected'. Unlike the filtered permeability described in the previous chapter, conventional thinking in the UK, reflected in national transport guidance, favours *unfiltered* permeability where pedestrians, cyclists and motor vehicles move in all directions along criss-crossing conventional streets. The 2007 guidance document *Manual for Streets* (quoted at the beginning of this chapter) implies, bizarrely, that street networks which allow cars to move freely in all directions will encourage people to walk and cycle. Like many ideas influencing official thinking in Britain, this one originated in the USA in a context where it was more understandable.

The traditional layout of North American streets is known as the 'gridiron' or rectilinear grid (Figure 8.7). In downtown areas of many American and Canadian cities, straight streets and avenues divide the built area into square or rectangular blocks of more or less regular sizes. More recent suburban developments abandoned this approach in favour of cul-de-sacs connected by distributor roads (Figure 8.8); these layouts increase distances between homes and neighbourhood centres or local facilities, often making them too far away to walk to (many of the suburban streets also have no sidewalks, creating shared spaces hostile to pedestrians). There are many other reasons why people in suburbs tend to drive more than people in downtown areas, so it is not surprising that urbanists and researchers began to associate permeable street patterns with more sustainable patterns of movement.

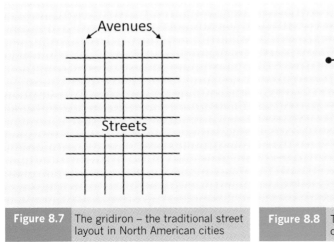

Figure 8.7 The gridiron – the traditional street layout in North American cities

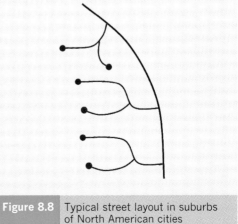

Figure 8.8 Typical street layout in suburbs of North American cities

Many studies have tried to measure the permeability of street networks – using block size, or numbers of junctions per square mile, for example. Some have also tried to control for other factors, like household income, household size or population density to see what difference street layout makes to travel behaviour. In North America, people do seem to walk more in neighbourhoods with permeable street layouts,[208] although the usual problem of what causes what has not been resolved. Part of the difference is due to self-selection: downtown areas attract people who prefer to walk anyway.

In Britain, where urban population densities are higher, the effect appears to be weaker. One recent study of four UK towns measured the permeability of neighbourhoods and related this to walking and cycling behaviour. It found that people walked slightly more in the most permeable areas but there was no significant difference in their cycling behaviour.[209]

Most of these studies have overlooked one fundamental point: the stylized layouts in Figure 8.7 and Figure 8.8 are both ways of designing streets for the car. The gridiron (or similar criss-crossing layouts with curved roads) shortens distances for walking but also makes every street a potential rat-run. In other words, it makes it easier to walk *and* easier to drive.

To make a more substantial difference, we would need to make the route we would take when walking or cycling more direct than the route we would take when driving. One group of American researchers identified a cluster of comparable areas with the two types of network shown in Figure 8.7 and Figure 8.8, plus one with filtered permeability. In the latter, barriers were placed across some of the roads in the traditional gridiron, removing through traffic but allowing pedestrians and cyclists through. Although it does not prove what caused what, people walked the most in those areas.[210] Two Dutch studies have likewise found that in areas where cycle routes are more direct than the alternatives for general traffic, rates of cycling are higher.[211]

This principle is controversial when applied to walking and cycling but generally accepted when applied to public transport – it is built into most of the models used by transport planners. If a guided busway enables a quicker and more convenient journey than the same journey made by car, projected bus use would rise. Building a new road alongside the busway would undermine the advantage offered by the bus, shifting some of those journeys back into cars. No transport planner would question these obvious implications. Why then have so many accepted the orthodoxy that shared streets are best for walking and cycling?

The UK guidance *Manual for Streets*, which recommends permeable streets open to all traffic, was influenced by the New Urbanist movement, which began in the USA and brought its influence to Britain through organizations like the Prince's Foundation. The aims of the New Urbanists are well meaning. The charter of the New Urbanism calls for neighbourhoods which are "compact, pedestrian friendly and mixed-use"[212] (although New Urbanist developments on both sides of the Atlantic have generally failed to break the mould of car dependency). The New Urbanists also support the move back to 'traditional streets' in North America, for reasons which are understandable but unhelpful when applied in a British or European context.

Several commentators on these issues have described their ideal situation as one where the car is "a guest in our streets".[213] But this is a guest that will never be house trained and will always pose a threat to humans who come into contact with it. In some circumstances, sharing of streets may be the least bad option, but to promote it as an ideal risks ruling out the alternative of streets entirely for people.

CHAPTER 9
"We are building too many flats"

Myths:
- "Land use planning has very little effect on how people travel".[214]
- We have been building too many flats – there is a shortage of family housing in Britain.
- "New settlements provide the opportunity and the economies of scale to truly fulfill the ambitions of sustainable development".[215]
- Locating jobs next to housing offers people the opportunity to work locally and reduces commuting distances.

Observations:
- Land use planning can, and sometimes does, exert a very powerful influence on how people travel.
- Only 19 per cent of households are families with a couple and one or more children.
- 64 per cent of British households have only one or two people.
- Flats make up only 20 per cent of the housing stock.
- New settlements built in Britain in recent years have generated very high levels of car use.
- The relationship between employment, housing and travel to work is complex: putting jobs next to housing doesn't necessarily reduce travel distances.

CHAPTER 9 We are building too many flats

This chapter is about land use planning and its relationship with transport. Most people involved in these two activities realize that each has a strong effect on the other. Even governments generally understand this, although they don't always behave as if they did. There is a belief (with increasing influence on governments in recent years) that argues that in effect you can't buck the market. It its most extreme form, writers from neoliberal thinktanks like Policy Exchange have argued that a state-run planning system should be scrapped in favour of voluntary arrangements between developers and "those who own neighbouring land".[216] Nick Boles (UK housing minister from 2012 to 2014) was the founding director of Policy Exchange, and Alex Morton, who articulated its position in a series of articles for *Planning Magazine*, was later employed as a planning adviser to the prime minister.

"Land use planning has very little effect on how people travel"

Leaving aside the politics and values, behind these views a laissez-faire planning policy would create serious problems for transport. In dense urban areas with well-distributed local facilities, people tend to walk more, use public transport more and drive less. Sprawling suburban housing estates and small settlements outside the main urban areas generate more traffic and longer journeys (although they may also distribute that traffic over a wider area). That much is quite easy to demonstrate. Figure 9.1 shows the relationship between population density and the amount of

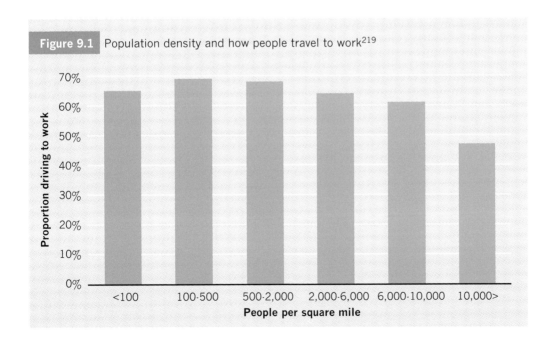

Figure 9.1 Population density and how people travel to work[219]

people who drive to work for local council wards in England and Wales from the 2011 census (the average size of a ward is around 5,500 people[217]). The only exception to the trend is the first category, the most sparsely populated rural areas, where local employment in agriculture or related work is more common (this only applies to commuting: people in rural areas drive the furthest in total[218]). The biggest effect – the smallest proportion driving – appears in the densest category, mainly the inner areas of the larger cities.

Many studies have shown similar relationships within and between cities in different countries. Although the relationship is clear, we come back to the same question raised in earlier chapters: what causes what? And does it follow that increasing population densities reduces travel by car? Some have argued that it works the other way round: people who prefer driving choose sprawling suburbs; people who prefer to walk choose denser urban neighbourhoods and the shape of cities is simply a reflection of those preferences. These questions have spawned hundreds, possibly thousands of studies and articles over recent decades and (in contrast to some of the debates discussed in previous chapters) some reasonably clear conclusions have emerged. Several of the more sophisticated studies have shown that personal preferences do not explain everything: the layout of neighbourhoods does affect how people travel, and can influence their transport and housing preferences.[220]

Housing density is not the only land use factor to influence travel, though it affects most of the others. In low-density areas, where people are more spread out, distances to services and public transport will be longer. In high-density areas, local shops, bus stops and railway stations will have larger catchment populations within walking distance. A study involving my UWE colleagues sheds some interesting light on how the location of facilities influences travel patterns. The SOLUTIONS project compared travel patterns in suburban neighbourhoods across four cities under pressure to expand.[221]

One of the neighbourhoods studied, Bradley Stoke on the edge of Bristol, has often been cited as a textbook example of how not to plan a new suburb. The plans were negotiated in the mid-1980s, when the Conservative-controlled Northavon council took a laissez-faire attitude to planning, rather like that recommended by Policy Exchange.[222] In place of the central planning of the post-Second World War New Towns programme, the plans for Bradley Stoke were agreed on a voluntary basis between individual landowners and developers.

The housing crash of the late 1980s depressed sales of the new housing and earned the town the epithet it took many years to shake off: 'Sadly Broke, negative equity capital of the UK'. In 2005, I interviewed residents, planners and councillors involved in its planning and early development.[223] Some of the residents told me how they had to fight for basic facilities like schools, shops, doctors' surgeries, even postboxes in the early days. Land allocated for a district centre lay derelict for over 20 years (Figure 9.2) until a shopping centre finally opened there in 2008.

The new centre had not yet opened when the SOLUTIONS researchers visited Bradley Stoke and 11 other suburban neighbourhoods around Bristol, Cambridge, London and Newcastle in 2007. They found very high levels of car ownership and use and low levels of walking in Bradley Stoke, as shown in Figure 9.3 (only a small proportion of journeys were made by bicycle in all the suburbs apart from those in Cambridge).

Figure 9.2 The author in 2005 on the site allocated for Bradley Stoke's district centre

Filton is an older, denser suburb with more local facilities. It is a couple of miles closer than Bradley Stoke to Bristol city centre. The survey area was near to a local high street. Filton had the highest share of walking and the lowest share of car use among the 12 areas – much higher than Bradley Stoke. So far, so predictable. But

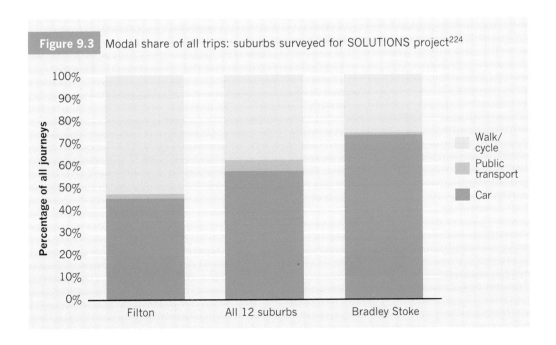

Figure 9.3 Modal share of all trips: suburbs surveyed for SOLUTIONS project[224]

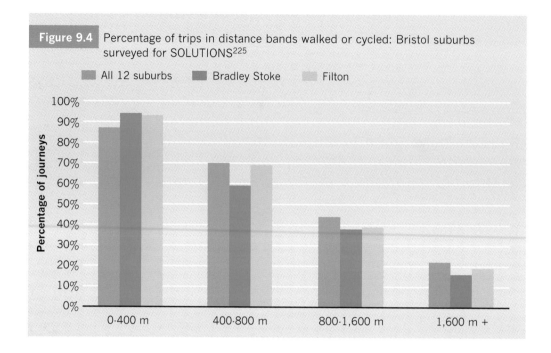

Figure 9.4 Percentage of trips in distance bands walked or cycled: Bristol suburbs surveyed for SOLUTIONS[225]

Figure 9.4 reveals something much more interesting. As we might expect, for shorter trips people are more likely to walk. As distance increases they are more likely to drive. Figure 9.3 shows that overall, people in Filton walk much more. However, Figure 9.4 shows that *for any given distance*, people in Bradley Stoke were almost as likely to walk as those in Filton. The main reason for the overall difference in Figure 9.3 was quite simple: Filton residents had more shops, services and facilities within walking distance. Bradley Stoke's facilities were further away, so more of these trips were made by car.

Findings like this have strengthened the case that planners can and should be able to influence how people travel, providing governments allow them to make decisions, instead of leaving everything to the market. Urban intensification, reducing walking distances and increasing catchment populations, can help to reduce car use and increase walking, cycling and public transport use. But like many policies implemented for good reasons it may also cause unintended consequences.

Figure 9.5 shows a similar relationship to Figure 9.1, although the units of measurement are different. It comes from a paper I wrote with two colleagues called 'The paradox of intensification'.[226] The figures are drawn from TRICS®, a database used by transport planners which shows how much traffic is generated by different types of new development. Basically, what it shows is this: denser dwellings generate less

CHAPTER 9 We are building too many flats

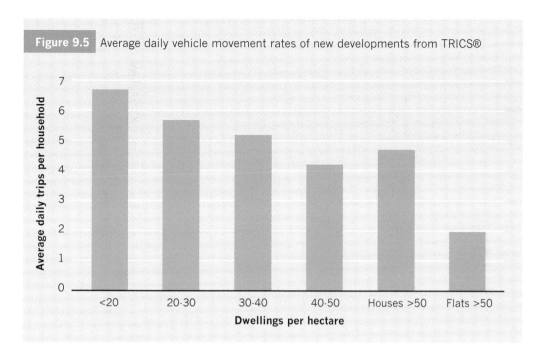

Figure 9.5 Average daily vehicle movement rates of new developments from TRICS®

traffic, but the difference is not proportional. So if you double the population density of an area you will reduce, but you won't halve, the traffic generated by each household. As a consequence traffic becomes more concentrated in intensified areas. This is what we mean by 'the paradox of intensification': those who drive least suffer the worst consequences of traffic, noise and pollution. It is one reason why people often object to high-density building in their 'back yard'.

The paradox of intensification

Urban intensification reduces overall car use but concentrates motor traffic in intensified areas, worsening their local environments.

There are five possible ways for planners and politicians to respond to this paradox:

1. Build fewer dwellings than we need and accept a rise in homelessness.
2. Carry on building at modest densities and accept a big loss of greenfield land.
3. Intensify and accept a worsening quality of life in the intensified areas.
4. Compromise with some deterioration in quality of life and some loss of green fields.

5. Intensify with more radical measures to restrain car use and improve the urban environment of intensified areas.

What these options might mean in practice will be discussed in Chapter 16.

"We have been building too many flats – there is a shortage of family housing"

In 2000, the UK Department of the Environment, Transport and the Regions under Labour Deputy Prime Minister John Prescott issued a guidance note, little noticed by the general public, which triggered some of the most radical changes in British cities of any government policy since the Second World War. The excitingly named *Planning Policy Guidance No. 3* (Housing) introduced minimum density guidance for new developments, maximum parking standards (no more than 1.5 spaces per dwelling) and a 'brownfield first' policy with targets. Although favourable housing market conditions offered an opportunity, the timing of the changes shown in Figure 9.6 leaves little doubt that the policy did directly contribute to a rapid rise in the density of new housing developments.

Much of this higher-density building came in the form of flats on brownfield sites in the inner areas of cities. For anyone who lived or spent time in British cities in the

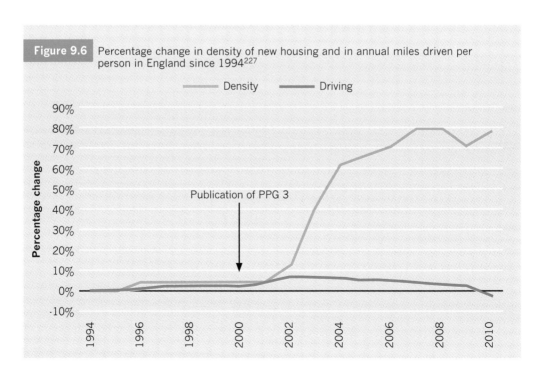

Figure 9.6 Percentage change in density of new housing and in annual miles driven per person in England since 1994[227]

Figure 9.7 Brindley Place, Birmingham

Figure 9.8 Harbourside, Hotwells, Bristol

early 21st century, the changes were unmistakable. The inner areas of many English cities increased populations rapidly from the late 1990s until the recession mainly due to high-density low-rise new apartments. While mistakes were made, and no doubt some horrors committed by greedy property developers, no one who knew places like Bristol's harbourside or Birmingham's canal district before and after redevelopment (Figures 9.7 and 9.8) could deny that a big improvement occurred.

Several studies have found these changes did not happen by accident or market forces alone: planning policy was the main reason. One clue to this is offered by the differences in planning policy between England and Scotland. The proportion of flats grew in England but not in Scotland, where planning policy placed less emphasis on increasing densities.[228] In England, average driving distances per person fell between 2002 and 2010 as shown on Figure 9.6, whereas the same measure rose in Scotland over the same period.[229] It would be impossible to determine *how much* the intensification policy in England contributed to the fall in car use but it did contribute, mainly because it increased population in places where people drive less, and constrained it in places where people drive more. (A chapter I have written in a multi-authored book, *Beyond Behaviour Change*, analyses this question in more detail.[230])

Transport is not the only reason to be concerned about the density of housing development. As discussed in Chapter 2, Britain has a serious housing shortage. The housing projections of 2010 estimated that there would be 5.8 million extra households in England by 2033, equivalent to a city nearly twice the size of London.[231]

Faced with these problems and growing hostility to greenfield building from its own supporters, the Coalition government in 2010 made a very strange decision to scrap national guidance on density and parking, and brownfield development targets. These changes were driven more by political considerations (including lobbying and payments from developers according to the *Daily Telegraph*[232]) than evidence about the impact of past policies. That said, the urban intensification under the previous government did create problems. Although they were cheaper than houses, flats built on high-value sites around city centres were expensive. The transport changes needed to support intensification were also fudged. The policy of parking restraint without parking control (discussed in Chapter 6) was doomed to failure, and the challenge posed by the paradox of intensification was neither recognized nor effectively addressed in many cities.

Another reason for the changes was 'localism' – a desire to devolve authority downwards, even though in other respects central government grip on local planning was tightening. (National planning guidance was changed to weaken the ability of local authorities to refuse planning permission; each authority was also required to maintain a five-year supply of land with permission for housebuilding.) In justifying the changes, the housing minister argued the previous guidance had:

> "resulted in developers building one or two bedroom executive flats, when the greatest need is often for affordable family homes".[233]

As I speculated at the time,

> "the Minister's suggestion that flat building has caused a national shortage of family housing was 'either a political ploy to play along with a popular misconception, or an early mistake on a steep learning curve'".[234]

If the minister's comment was an early mistake, there has been little sign of any correction. The density figures in Figure 9.6 were only available up to 2011. More recent statistics are available on the mix of newly built housing, which confirm that the expected change did occur. In 2008-9, 50 per cent of new dwellings in England were flats; by 2013-14 this had fallen to just 29 per cent. At the same time, houses with four bedrooms or more rose to 26 per cent of new dwellings.[235] The Coalition government did not appear to appreciate the full implications of these changes. Where most building takes place on brownfield land, and greenfield building takes the overspill that brown fields cannot accommodate, even small changes in density can hugely increase the loss of greenfield land, as illustrated in Figure 9.9.

In the simplified example of Figure 9.9, roads, schools and everything which accompanies housing have all been ignored, but it illustrates the basic point: small differences in density can make a much bigger difference to the amount of greenfield land required for housing.

Illustration

Imagine two possible scenarios for new residential development. The ratio of four brownfield dwellings to each greenfield dwelling was the situation in 2009.[236] In other respects, the numbers are purely illustrative:

High-density scenario:
- 4,000 dwellings are built on 40 hectares of brownfield land.
- 1,000 greenfield dwellings are built on **25** hectares of greenfield land (a density of 40 dwellings per hectare).

Low-density scenario (brownfield and greenfield densities are both halved):
- The brownfield land can now only support 2,000 dwellings.
- The remaining 3,000 dwellings must be built on greenfield land.
- The density of greenfield building falls to 20 dwellings per hectare.
- Those 3,000 dwellings will now need **150** hectares of greenfield land (3,000 divided by 20).

So by halving the density of development, we need **six times** as much greenfield land to accommodate the same number of dwellings.

Figure 9.9 Influence of Housing Density on the Loss of Greenfield Land

The proportion of flats did certainly rise under the previous Labour government; the 50 per cent of new dwellings reached in 2008/9 was the high point. But was it true and is it true now that we have too many flats, and too little 'family housing'?

Question: What percentage of households in England and Wales are made up of families with a couple and one or more dependent children?

Over the years, I have posed this question to many planners, developers, councillors and even to a housing market analyst that I met by chance in a bar (yes, it does happen sometimes). The vast majority of people that I ask over estimate: the correct answer, illustrated in Figure 9.10, is only 19 per cent (including single parent families and other households with dependent children brings the total to just 29 per cent).

Question: What percentage of England's housing stock is flats?

The answer is 20 per cent. As illustrated in Figure 9.11, this is considerably lower than Scotland (38 per cent) or any of the other large European countries – all of which have a lower population density than England.

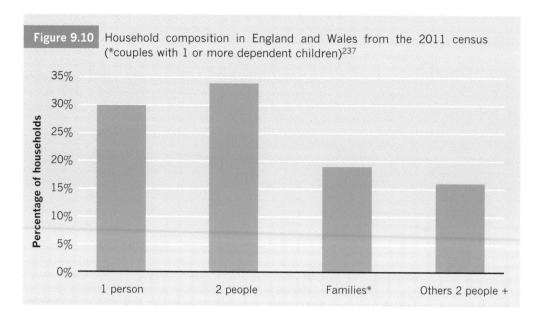

Figure 9.10 Household composition in England and Wales from the 2011 census (*couples with 1 or more dependent children)[237]

Even at the height of the boom, new development was adding less than 1 per cent to the housing stock each year, so the flat building of the early 2000s made very little difference to the overall balance of the national housing stock. If half of the 5.8 million dwellings the government was projecting in 2010 were built as flats, this would still leave three-quarters of England's dwellings as houses.[239] It was, and is, only a slight oversimplification to say England is a country of single people and couples

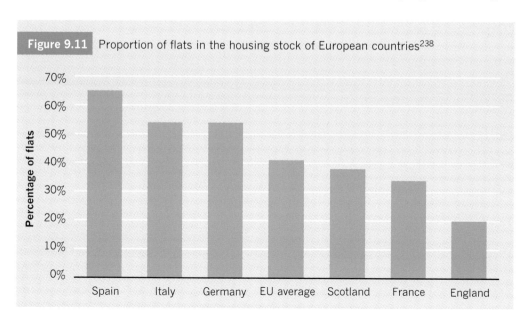

Figure 9.11 Proportion of flats in the housing stock of European countries[238]

living in houses designed for families. But of course 'shortage of family housing' will always make a better headline than 'shortage of housing for single people' – even when we see some of those single people sleeping on the streets.

When I have presented this analysis to people who work in planning or housing, a common response is that it may not be a national problem, but there is a serious problem in some areas. Even this perception is highly distorted. Across England and Wales there are just 150 local authority wards, mainly in inner urban areas, where more than 75 per cent of the dwellings are flats. By comparison, there are 4,021 wards where over 90 per cent of the dwellings are houses (nearly half of the 8,570 wards in England and Wales).[240]

Many local planning documents refer to shortages of family housing.[241] Why is this? Apart from the media and distorted perceptions of household size, local authorities are also influenced by their responsibility for social housing. The 'right to buy' legislation introduced in the 1980s, that gave social housing tenants the right to buy their homes, has been compared to "filling the bath with the plug pulled out".[242] While the Scottish government has put the plug back in, and the Welsh government has given local authorities the power to suspend the right to buy in areas of housing shortage, English legislation has extended those rights several times despite ever-lengthening waiting lists. Council houses have proved a more attractive buy than council flats. Larger families on low incomes are always more likely to need housing assistance, whereas most single people are not even allowed on to the waiting lists in the first place.[243] While there are shortages of *social* housing for larger families in many areas, these factors have distorted the perceptions of planners responsible for *all* housing, most of which is privately built and occupied. Ironically, social housing providers are still building more flats than private developers. In the private market nearly a third of all new dwellings are now houses with four bedrooms or more.[235]

"New settlements are a sustainable form of development"

At times of rising population and housing shortage, UK governments of different political persuasions look again at the idea of building new towns. The New Towns Act of 1946 provided the framework for 32 New Towns[244] designated between 1946 and 1970. The New Towns built under this act were planned by development agencies set up by the national government, which compulsorily purchased land at 'existing use' prices – usually the market price for agricultural land. The difference between these two can be very substantial. In 2013, the average price for agricultural land in the UK was £8,500 per acre,[245] whereas the average value of greenfield land once developed ranged from just under £300,000 per acre in Scotland up to £1.5m per acre in south-east England.[246]

The New Town development agencies directly built some of the infrastructure (such as roads) within the New Towns and worked with local authorities to provide services such as schools. They were able to recoup their initial investment through the sale or lease of land, which grew in value as the towns were built.

The New Towns Act remains on the statute book, but it has not been used to designate any more New Towns since 1970. However, some smaller new settlements have been initiated by private developers working with local authorities. In 2007, Prime Minister Gordon Brown announced a competition to find 10 sites for 'eco-towns', designed to higher environmental standards.

By 2007, the political climate had changed; the New Towns Act came from a socialist tradition of central planning, which governments of all parties wanted to avoid. Eco-towns would be built by private developers with profits on the sale of the land going to the original landowners. The announcement of the eco-towns programme provoked widespread opposition in places where promoters were bidding to build them. The government responded to public and media criticism with assurances that eco-towns would go through the normal planning process, requiring permission to be granted by local authorities.

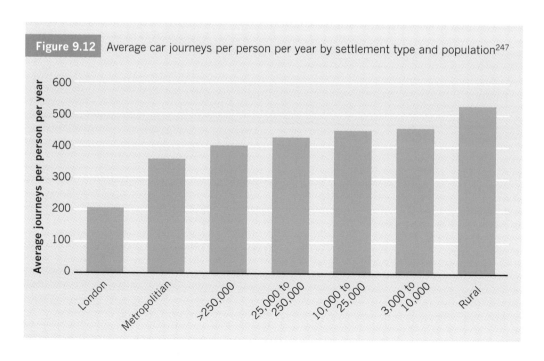

Figure 9.12 Average car journeys per person per year by settlement type and population[247]

CHAPTER 9 We are building too many flats

No local authority was likely to agree to a large new town in its area, and private developers were unlikely to take the risk of building one, particularly if they had to pay high land prices, so eco-towns would be relatively small: the planning guidance specified a minimum of 5,000 dwellings.[248] As sustainable transport was one objective of the programme, the small size of the planned eco-towns raised a problem. Figure 9.12 shows a very clear pattern that smaller towns and villages generate more traffic per inhabitant than larger towns and cities.

The few privately developed new settlements started after 1970 have proved particularly car dependent; Table 9.1 lists five of them. There is no agreed definition of a new settlement (as distinct from the 'New Towns' designated under the New Towns Act), nor, as I found when talking to civil servants a few years ago, does anyone in government keep a comprehensive list of them. Table 9.1 includes only 'freestanding' settlements; it does not include the much larger number of 'urban extensions' (new suburbs such as Bradley Stoke), which are physically attached to a larger town or city. Table 9.1 illustrates all the freestanding new settlements built in Britain since the 1970s that I am aware of, or could find reference to in the literature, and which were occupied at the time of the 2011 census. Two of them, Ivybridge and South Woodham Ferrers, were expanded around existing villages from the 1970s onwards. At what point a growing village may be considered a new settlement is a matter of judgment. Wixams was at an early stage in its development in 2011; several more new settlements – not listed – were started later, or are planned.

All five settlements in Table 9.1 are near to, but separated from, a larger town or city. This partly explains the striking pattern of the last two columns: average commuting distances from these settlements are all much higher than the national average, and the vast majority of commuters drive. Three of the settlements have railway stations,

Table 9.1 Patterns of commuting in free-standing new settlements[249] (*expanded around existing villages)

New settlement	Near to	Begun	Population	% Commuting by driving	Average commuting distance (km)
Ivybridge	Plymouth	1970s*	11,851	74.5%	16.9
Wixams	Bedford	2007	976	81.8%	16.9
Cambourne	Cambridge	1998	8,186	76.1%	19.0
Dickens Heath	Solihull	1997	3,992	82.5%	16.1
South Woodham Ferrers	Chelmsford	1970s*	16,453	67.2%	17.2
England average:				60.2%	12.1

but rail is only significant in one of them: 14 per cent commute by rail from South Woodham Ferrers, Essex, which is just under an hour from London.

In 2008, I briefly became an unpaid government adviser, contributing to guidance documents on transport in eco-towns written by the DfT[250] and by the Town and Country Planning Association (TCPA) for the Department for Communities and Local Government (DCLG).[251] The TCPA worksheet, which is still available on its website, was one of the most radical transport documents ever endorsed by a Westminster government. It incorporates several of the principles in this book, including European-style carfree development and filtered permeability. At the same time, I was informally coordinating several national environmental groups who were concerned these principles would fail if eco-towns were built in the wrong places. Access to good rail services would be absolutely critical (which is a necessary but not a sufficient condition for sustainable transport), we argued.[252] With hindsight, we could also have made a stronger argument about settlement size: it is much easier to plan for sustainable transport in one large new town or city than it is in many small new towns or villages.

Opposition from local authorities and the general public weakened the government's commitment to the eco-towns programme as the general election approached. Four sites were chosen for new suburbs rather than new towns. Only one of them had access to rail. The opposition parties condemned the whole idea and the incoming Coalition government effectively buried the programme. Four years later, however, they returned to a similar idea, but without the 'eco' element. Their 'Prospectus for locally led garden cities' had little to say about transport, but was explicit on one point: where rail connections are poor, buses can be used instead.[253] For the reasons discussed in Chapters 5 and 6, this strategy is unlikely to support a sustainable pattern of travel: rail is particularly important for people without cars in smaller settlements and buses are the least likely form of public transport to attract drivers out of their cars. The prospectus recommends a minimum size of a 'garden city' of 15,000 homes – larger than the eco-towns but still towards the smaller end of the settlement bands in Figure 9.12. It says nothing directly about housing densities, but the original concept of the garden city was based around two-storey family housing at relatively low densities. That made some sense in the early 20th century, when household sizes were much larger, but the organizations promoting the idea today, including the TCPA, still seem to favour that type of development,[254] when most households have just one or two people.

Like the eco-towns programme, the garden cities prospectus also fails to explain how new infrastructure will be funded (except to say that a "private sector funding solution" must be found). If private developers have to pay market prices for the land, then whatever mechanism is used, the funds available for new infrastructure will never match those available to the new town development corporations.

The recent experience of new settlements has revealed another problem, which reinforces their unsustainable nature. Risk-averse private developers and local authorities fearful of public opinion tend to think small when designing new settlements. Prettily drawn master plans aim to reassure the public that the new settlement will be neatly self-contained and not too threatening.[255] But in future years, surrounding landowners and planning inspectors are likely to view the growing new settlement as a prime candidate for more housing than originally planned; this has already occurred in the case of Cranbrook in east Devon, where building began in 2011.[256] Over time, this process tends to produce a medium-sized town with housing estates spreading away from a town centre originally designed for a much smaller settlement. Most older towns and cities have ex-industrial land or low-quality old housing around their centres, offering opportunities for redevelopment as they grow. Not so in small new settlements – once the centre of a new settlement is surrounded by new housing, it will be very difficult to expand it later on. This has been the experience of towns like Ivybridge and history seems about to repeat itself in the current generation of new settlements.

In the run-up to the 2015 general election, the Labour party commissioned its own study of housing policy (the Lyons Review). Their final report repeated some of the myths about flats and family housing and, with a few minor differences, endorsed the Coalition's policy of building small unsustainable new settlements.[256i]

"Locating jobs next to housing reduces commuting"

Some planning authorities believe that unsustainable commuting patterns in new settlements and suburbs can be solved by putting more employment opportunities within them, but the evidence supporting this view is rather weak. Some years ago, I was involved in several 'examinations in public' into local and regional plans covering south Devon. The early drafts of a 'Local Plan for the South Hams District' proposed an extension to a housing estate on the edge of Plymouth. It bordered a large employment area, which was also planned for expansion. The planners followed the conventional wisdom of the time (which still has some influence today) arguing that this location "could be considered sustainable in strategic transport terms given that it is located relatively near to an existing employment area".[257] I analysed the census statistics for the existing housing estate (Newnham) and found that the journeys of many residents driving to work were much longer than the average for the city. Despite having many jobs on their doorstep, only 2 per cent of the working people on the estate walked to work nearby.[258] This is not unusual: Bradley Stoke also has a lot of local employment but relatively little local working. Part of the reason is that it is near two motorways, giving the opportunity to drive longer distances; it is generally true that people living near major roads tend to travel further to work.[259]

Patterns of living and working are generally diverse and many of the stereotypes are misleading. Commuting only accounts for one trip in six; most travel is for shopping or 'personal business' of various kinds.[260] If that surprises you, bear in mind that full-time employees are only around a third of the population.[261] Commuting into London from the home counties is much more limited than most people imagine, ranging from one in six of the working population of Essex to just one in ten of Kent and Surrey.[262]

Ivybridge, in south Devon, is one of the new towns listed in Table 9.1. About 5 miles of countryside and creeping development separate it from Plymouth, which is 20 times bigger. The rest of the surrounding area is rural with a few smaller towns and many villages. At the 'Examination in Public' into the Regional Spatial Strategy in 2007, a representative of Ivybridge town council said the town really needed more local jobs, so people wouldn't have to commute into Plymouth; this was, and still is, a common perception.

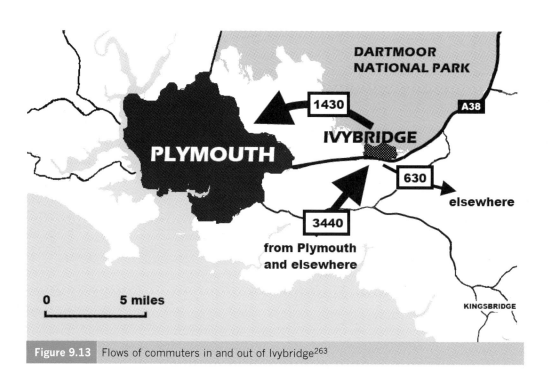

Figure 9.13 Flows of commuters in and out of Ivybridge[263]

But a report commissioned for the regional assembly told a rather different story – one which remains relevant for small settlements on the edge of big cities. In reality, Ivybridge had a big surplus of jobs: 5,000 jobs for a working population of 3,650. Plymouth also had a (relatively smaller) surplus, but whereas 90 per cent of working Plymothians worked in the city, fewer than half the jobs in Ivybridge were filled by local people.

Figure 9.13 shows a pattern of two-way commuting. Every job created in Ivybridge was more than twice as likely to be filled by an inward commuter than by a local person. With the A38 Expressway linking Ivybridge and Plymouth, 80 per cent of commuting out of Ivybridge was by car. A railway station that opened in the 1990s made little difference. Simplifying a little, it seems that wherever you put jobs, people will travel to them from all directions. Wherever you put housing, people will travel outwards to work in all directions.

A common mistake made by planners and many other people considering these problems is to assume that decisions about who works where are made by employees. In reality, such decisions are mostly made by employers, from a pool of candidates travelling from all directions. The most important decision made by candidates is the not the positive "Where would I like to work?" but the more pragmatic "Where might I be willing to work?" Also, most people don't want to live right next to work: they value some transition time between work and home.[264]

Although commuting does not generate many trips on its own, it is important because of its peak-time concentration and because it influences where people live and many of their other journeys. Does the analysis in this chapter imply that trying to influence how people commute through planning is a waste of time? Not necessarily. Studies in this country[265] and elsewhere[266] suggest the balance of jobs to working people in an area does influence travel distances, but the pattern is more complex and the strategy needs to be more sophisticated than simply putting jobs next to housing, as discussed in Part II.

CHAPTER 10
Summary: myths, values and challenges

Several of the myths discussed in Part I circulate within the transport world, promoting conventional wisdom or defending vested interests. Others, more likely to be heard among the general public, argue that we cannot, or should not, act to restrain driving, flying or vehicle ownership. You may have formed your own view about whether all of them can be dismissed as urban myths, or whether any of them encapsulate elements of the truth.

UK governments abandoned any serious attempt to restrain car traffic some time ago, although the belief that they are trying to 'get people out of their cars' seems to endure regardless. Underlying both the myths and the attitude of government is a set of values, opposed to the ones I listed at the beginning of this book, that regard environmental sustainability as a 'nice to have', but only if it does not conflict with freedom to travel or increasing consumption of goods. It is sometimes argued that there is no conflict between those aspirations,[267] or that economic growth will move us in the right direction with no need for restraint.[268] The discussion so far has suggested reasons for doubting that view, at least as applied to transport. Road building encourages more traffic (whether it benefits the economy or not); expanding aviation capacity threatens our ability to meet climate change targets. Electric cars will help, although they create other problems and offer no panacea. Public transport improvements on their own make little difference to traffic or pollution, whereas constraints on car ownership and parking, like them or not, can have a direct impact.

If 'restraint' appears to imply life getting worse, there are ways that restraining traffic can improve people's quality of life, particularly in dense urban areas, where more of us are likely to be living in future. Those are some of the challenges. Part II will look at potential solutions, starting with some proposed over 50 years ago.

PART 2
Sustainable solutions

CHAPTER 11
Four options for traffic in towns

In 1960, the transport minister Ernest Marples set up a committee to write the report 'Traffic in towns', which became an unlikely paperback success.[269] The Buchanan report, as it became known was written at a time of transition, and of expectation. It began with an uncannily prescient forecast of future car ownership (see Figure 11.1) and its consequences.

This "rising tide of cars", it predicted, "will not put a stop to itself until it has almost put a stop to the traffic." As a majority of voters would soon become car owners, politicians would "be anxious to please the motorist and frightened of annoying him" making "drastic action politically difficult".

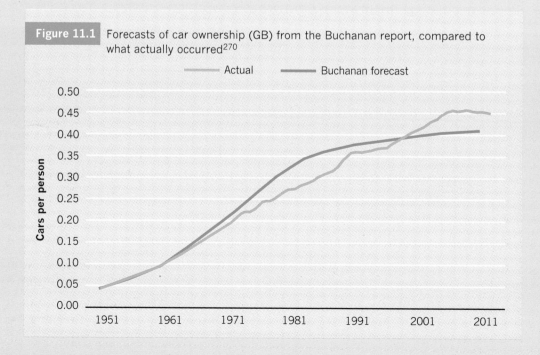

Figure 11.1 Forecasts of car ownership (GB) from the Buchanan report, compared to what actually occurred[270]

Many transport commentators in recent years have blamed the Buchanan report for leading the UK in a wrong direction, one which later generations would repudiate and partially dismantle at considerable cost. In one important respect – its call for large-scale road building including urban motorways – that criticism was justified. But the report itself was more sophisticated than its more recent caricatures. Its analysis of the underlying problem and the choices we face in urban areas remains relevant today.

As car ownership grew, so would the conflict between vehicular access and the urban environment (including the freedom to move around on foot). Bypasses around towns would remove only a small proportion of through traffic; most of the problem was generated by vehicle movements within towns and cities. Drawing an analogy with a hospital, the problem could be alleviated by dividing each town into 'environmental areas', where people live and work (the rooms) serviced by a network of distributor roads (the corridors).

'Extraneous traffic' was to be removed from environmental areas by closing roads around the boundaries to through traffic: vehicles would drive in and out through a limited number of access points. Some areas would be selectively pedestrianized, but the report took pains to point out that the policy was about much more than the 'pedestrian precincts' which became associated with the Buchanan approach.

Where roads were closed, or streets pedestrianized, pedestrians and cyclists would be able to move in all those directions where motor vehicles were blocked – the filtered permeability discussed in Chapter 7. This principle was never properly implemented in Britain and the concept of the environmental area gradually fell out of favour. However, many European cities took it and developed it into better solutions than Buchanan ever imagined. We examine some of these in Chapter 12, and one of the most radical experiments, directly influenced by Buchanan, in Chapter 13.

Within the environmental areas, Buchanan argued:

> "it might be sufficient merely to exclude all extraneous vehicles, but even the area's own traffic might increase beyond the [environmentally acceptable] limit as a result of the conversion of houses to flats or… high car ownership".

In those circumstances, planners faced three options:

1. Rearrange and rebuild the town to accommodate more traffic.
2. Restrain vehicular access.
3. Accept the consequences of congestion and a degraded urban environment.

Elsewhere in the report, a fourth option is mentioned, though mainly to downplay its potential:

4. Voluntary behaviour change, reducing car use or changing its distribution.

Environmental capacity, restraint and behaviour change

Options 2, 3 and 4 were mainly discussed as context for the central thrust of the report, which is all about the rearrangement and rebuilding of towns. The report introduced a new concept of 'environmental capacity' – a maximum volume of traffic that a street or network of streets could accommodate without imposing unacceptable conditions on pedestrians and occupants of buildings. Appendices to the report suggested some ways this capacity might be calculated. If the environmental capacity of an area was exceeded, and if this was considered unacceptable, then either restraint or rebuilding would be necessary.

The concluding chapter mentions four possibilities for traffic restraint: "a series of permits or licences... to control the entry to certain defined zones", electronic road pricing (a novel idea at the time), parking policy and subsidizing public transport to offer an advantage over the use of cars.

Subsidising public transport is really more of a voluntary behaviour change measure; Buchanan did not place much emphasis on the potential for traffic reduction through such means. The only other voluntary measure mentioned in the report was the staggering of hours of work to reduce pressure at peak times. Working hours and places of work have become more flexible over the years, although the basic problem of peak-time travel has not changed.

Rebuilding towns to accommodate more traffic

The report argued that urban areas could and should be converted to accommodate more traffic by demolition, road expansion and grade separation: flyovers, underpasses and pedestrian walkways above the traffic such as the one in Figure 11.2.

The disadvantages of structures like that in Figure 11.2, particularly for less mobile pedestrians, are clear. Apart from steep concrete steps, that particular example used to have an escalator at each end, which could help people with some disabilities but were not much use to wheelchair users; they cost money to maintain, broke down from time to time and were eventually abandoned. These disadvantages seem so obvious that it is difficult to imagine today how planners and local politicians went along with the idea.

Figure 11.2 Pedestrian walkway over Bristol's inner ring road

The combination of pedestrian walkways or underpasses and road expansion at ground level was repudiated by planners in later years, partly because of the impact on quality of urban life, but also because it proved less effective than Buchanan envisaged. Every journey begins and ends somewhere, usually in two buildings. If vehicles access the buildings on one level – usually ground level – then capacity will be limited at the point where the vehicles meet the buildings. Moving pedestrians out of the way makes little difference when the competition for space is dominated by motor vehicles. Multi-level roads and multi-storey car parking can only solve the basic problem if roads and buildings are both designed to allow vehicular access at multiple levels. Buchanan used the image of the Penn Center in Philadelphia to illustrate the potential for this (Figure 11.3). Note the different levels on which pedestrians and vehicles of different kinds can access the buildings.

Multi-level vehicular access has been tried in many places around the world since then (think of major airport terminals, for example), but it was never going to provide a general solution to the problem of traffic in towns. Elevated roads damage urban environments and reduce neighbouring property values; they are also expensive to

Figure 11.3   Penn Center, Philadelphia, taken from 'Traffic in towns' (Buchanan report, 1961)

build and maintain. Burying roads may have advantages in some circumstances but will always be constrained by cost. Land values in west London may eventually help to finance the burial of the Hammersmith Flyover, but the same solution will never be feasible in the 'average' urban area. So if multi-level access to buildings is not feasible, the only other way to accommodate much larger volumes of urban traffic is to knock down and spread out. The problem is not just where to fit the roads, but also the traffic generated by the buildings and the activities within them.

The road network around Bristol city centre illustrates the critical flaw in Buchanan's preferred approach. In the 1960s, Bristol council published plans to build inner and outer ring roads, demolishing buildings where necessary and filling in parts of the city's floating harbour. Public protests eventually halted the plans, but some buildings were demolished to build a partial inner ring road with six or eight lanes of traffic (Figure 7.6, page 67) and pedestrian walkways crossing at a higher level (Figure 11.2). More buildings were demolished to build a motorway, the M32, which connects the inner ring road to the national motorway network. Figure 11.4 was taken from another walkway, over the M32 in the position shown on Figure 11.5, looking away from the city centre. The traffic jam, which recedes into the distance, is typical of the in-bound congestion at that end of the motorway in the morning rush hour. Buses are forced to wait in the traffic jam before they can reach the bus lane.

Most of that in-bound traffic is heading for destinations within the central areas of the city, so widening the motorway or the ring road might move congestion from one place to another, but to significantly reduce it the number of vehicles heading into

CHAPTER 11 Four options for traffic in towns

Figure 11.4 Traffic jam on the M32 heading towards Bristol city centre in the morning rush hour

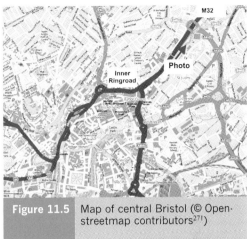

Figure 11.5 Map of central Bristol (© Openstreetmap contributors[271])

central Bristol would also have to be reduced. The relatively few buildings demolished to make way for the motorway and the ring road were not enough to reduce the intensity of trip-generating activities in central Bristol. Bristol has the most congested urban main roads of any British city apart from inner London,[272] due partly to the large number of business and commercial parking spaces in the central areas (20,000) in addition to residential, public and on-street parking.[273]

If the authorities in Bristol, or any other city, wanted to avoid traffic restraint and solve the problem of urban congestion through more road building, they would need to reduce the intensity of trip-making activities – the jobs per square mile in employment areas and residents per square mile in residential and mixed-use areas. This is the 'spread out' option, which had already occurred in many North American cities when Buchanan was writing, and was about to occur in the second wave of New Towns in Britain. Milton Keynes, for example, was planned in the 1960s at low densities around a network of 'grid roads' with a 70 mph speed limit (Figure 11.6). If we were willing to knock down enough buildings (including listed buildings), we could make our existing cities more like Milton Keynes, providing, of course, we allocate enough green fields to accommodate the people and businesses displaced. (These would be additional to the green fields needed for new housing, discussed in Chapter 9.)

Similar considerations apply to the capacity of inter-urban roads: the vast majority of journeys begin and/or end in urban areas, so the capacity for inter-urban travel will be constrained by the road capacity of towns and cities.

If voters and politicians are unwilling to accept bulldozing or effective restraint then congestion will impose restraint by default. Buchanan believed the electoral power

Figure 11.6 Milton Keynes: urban grid road with 70 mph speed limit

of the motorist would make this politically impossible, but later generations have proved more tolerant of congestion than he imagined. As building our way out of congestion has proved impractical and the alternatives include things like road pricing or higher taxes, a lot of motorists seem to prefer sitting in traffic jams. Others imagine the jams would disappear if only the council would do things like removing bus lanes and making it easier to park. In reality, more parking, if it is used, generates more traffic. And increasing the capacity of some roads causes more congestion on others, and at junctions, unless it is part of a plan to 'knock down and spread out'.

New technology and the situation today

If we compare the situation Buchanan described with the one we face today, none of the fundamentals have changed. For as long as vehicles controlled by individuals remain the dominant form of transport, society will face those same four options. Technological developments aimed at changing the way cars are driven (eg computer coordination of traffic signals and GPS route finding) are simply different forms of restraint or voluntary behaviour change. Jet packs and vertical take-off aeroplanes, though technically possible, remain as irrelevant today as they were when Buchanan speculated and dismissed them as feasible alternatives.

Electric (or hydrogen) vehicles will change the likelihood and the consequences of road congestion. Electric cars are much cheaper to use than petrol or diesel cars. A mass market switch would deprive governments of the revenue from fuel taxes unless they decide to impose either a selective tax on electricity (difficult but not impossible) or a national road pricing scheme. Neither is likely to be popular.

Meanwhile, the UK government has placed considerable faith in the potential for 'driverless cars' to reduce congestion and CO_2 emissions.[274] The term 'driverless cars' is misleading (researchers prefer 'autonomous vehicles') since the technologies can be applied to other types of vehicle and the transition to entirely automatic control for private cars will pass through several intermediate stages. Some automatic functions such as cruise control have been available for some time. Others such as anti-collision and parallel parking technology have been introduced on mass-produced cars more recently. Cars that can run safely on public roads with no driver, in circumstances including fog and heavy rain, are likely to take several decades to reach commercial viability and, like electric cars, they will be very expensive at first. Autonomous road vehicles have the potential to reduce congestion because they can use road space more efficiently, by running closer together at coordinated speeds and interacting more swiftly at junctions. Automated dropping-off followed by driverless parking may also increase the capacity of buildings to receive incoming journeys.

Driverless electric vehicles following narrower guided routes could free space for pedestrians, cyclists and the public realm. They could also provide better forms of public transport. Personal rapid transit systems like Ultra at Heathrow's Terminal 5 can be programmed to carry a passenger directly to any stop on a network (Figure 11.7). Unlike buses or trams, there is no need to change. Although they run quite slowly (up to 25 mph) they are faster than buses for most journeys because they don't stop so often. They also consume much less energy than other motor vehicles. At full power, a single tram carriage consumes around 750 times the electricity of an Ultra pod.

The Ultra pods are small, carrying up to four passengers, like a taxi (Figure 11.8). The first personal rapid transit system has been running since the 1970s, between a university and town centre in West Virginia. Those vehicles are larger, providing something more like a minibus but with the same point-to-point service. Both of these systems run on guideways but if driverless electric vehicles can be designed to run on public roads, they could also replace buses and taxis, though not high-volume urban rail systems.

Studies into the potential for autonomous vehicles suggest that the big reductions in congestion will only materialize when the vast majority of vehicles on the road are moving autonomously. If many drivers insist on overriding the controls, to change speeds or route choices, then the gains will be much more modest. Hence these choices bring us back to the options of restraint or voluntary behaviour change.

Figure 11.7 Ultra, personal rapid transit system at Heathrow Terminal 5

Figure 11.8 Inside an Ultra pod

Learning from international good practice

The next four chapters look at specific examples of measures which have succeeded, to varying extents, in reducing car use and improving the urban environments of British and European cities. These all concern transport in urban areas; the countries which provide some of the best examples have not performed much better than the UK at the national level. Germany, France and particularly the Netherlands have built dense motorway networks which undermine any attempts to promote sustainable inter-urban transport. As a result, national travel patterns across Europe are remarkably similar, particularly when compared by distance travelled rather than number of trips, as illustrated in Figure 11.9; investment in high-speed rail, particularly in France, has made little difference to the overall pattern.

The word 'sustainable' needs some clarification. It has become one of the most abused terms in the English language. The most commonly accepted definition of 'sustainable development' is the one proposed by the Brundtland Commission in 1987:

> "development that meets the needs of the present without compromising the ability of future generations to meet their own needs".[275]

This rather open-ended definition begs the question in terms of the timescales we are talking about. Can we consider an activity sustainable if it continues for tens, for hundreds, or thousands of years without affecting future generations?

UK governments have redefined sustainable development to include "high and stable levels of economic growth".[276] This concept is difficult to reconcile with the Brundtland definition unless we ignore the longer term or assume that some sort of

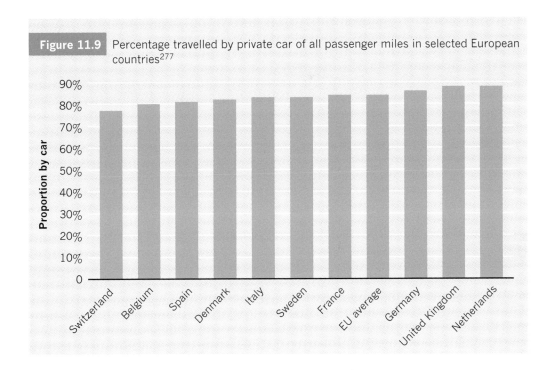

Figure 11.9 Percentage travelled by private car of all passenger miles in selected European countries[277]

miraculous change in manufacturing processes and waste disposal is just around the corner.

If we took the Brundtland definition literally, and did not limit our time frames, then it would allow us to do only two things:

1. use resources that are 100 per cent renewed or reused and;
2. generate only waste that is reabsorbed by the Earth faster than humans are creating it.

Clearly we are a long way from such a scenario at the moment. So where I use the term 'sustainable transport' in the following chapters it is a tendency statement – a movement towards sustainability by reducing modes with the greatest environmental impact – particularly the car, but also aviation and heavy goods vehicles (HGVs).

Learning from Europe has been common practice in transport and planning across the English-speaking world. On measures like modal share and quality of public spaces, cities in northern Europe are generally more sustainable than British (and particularly American or Australian) equivalents, although there are some exceptions, reviewed in Chapters 14 and 15. Some of the European cities and towns discussed in Chapters 12 and 13 are often used as illustrations of best practice, although not always in helpful ways. If you read some of the case studies on www.eltis.eu, for example, you will find a lot of positivity and very little discussion of problems or unintended consequences. This approach to international lesson-learning has encouraged the myth-making described in Part I. Of course they can do those things in all those other countries, where they have wide roads, the sun always shines and very little ever seems to go wrong – it's not like that over here.

In the summer of 2006 I set out on what became the first of several bike rides across seven European countries, initially to study the carfree developments described in Chapter 13. Along the way my interest strayed towards some broader questions. Which cities have succeeded in taming motor traffic and improving the quality of urban life? How did they do it, and what can we learn from them? In sharing what I found with people back in Britain I started to encounter some of the myths (debunked in Part I) about cultural differences on the other side of the Channel. There *are* institutional differences between countries, particularly in the powers of city authorities, but a visitor seeing these and British cities for the first time, with their transport problems and their range of solutions, would be more struck by the similarities than the differences.

The three cities I describe in the next chapter – Freiburg, Groningen and Lyon – have been chosen because they have significantly reduced car use and because each of them exemplifies the approach taken by more progressive German, Dutch and French cities respectively. Chapter 13 will describe some more radical experiments with carfree development, and some international trends in carfree living. Chapters 14 and 15 will describe the experience of three British cities: London, Brighton, and Cambridge, which have also made significant progress in recent years.

One consequence of the economic obsession discussed in Chapter 4 has been a proliferation of attempts to quantify the impacts of policies and programmes. Despite a sophisticated industry behind them, transport models still struggle to predict the behaviour of unpredictable human beings.[278] Trying to explain what happened in the past is a little easier, but not much. It may be possible to estimate the effects of specific measures like the London congestion charge, though as we shall see in Chapter 14, even those remain contested. About the causes and effects of broader longer-term changes across different cities and countries, certainty will always elude us.

The next four chapters will start with a simpler aim, to describe who did what, and what changed. In looking for possible explanations I will draw on secondary sources, direct observation and a mixture of interviews with transport professionals, political leaders and community activists in each of the cities. As we piece together these accounts, several patterns emerge: one of these is that plans in all of the countries never run quite as smoothly as planners intend.

CHAPTER 12

European cities: inspiration and similarities

This chapter will look at three cities, Freiburg, Groningen and Lyon, chosen because each of them provides a good example of transport practice in Germany, the Netherlands and France respectively. We will look at the recent history of transport planning in each city and the effects on travel behaviour. The final section will draw comparisons and some tentative conclusions from the experience of all three.

Freiburg, Germany[279]

My first destination in 2006 was Freiburg, a city in south-west Germany with a population of around 230,000 (about the same size as Southampton in the UK). Surrounded by the Black Forest and the wine growing area of the Rhine valley, it has acquired a reputation as Germany's green capital: a centre for the solar industry and a wealthy, growing city where people move for a better quality of life. It wasn't always like this. In the 1950s and 1960s, it was a conservative provincial town with a growing traffic problem, which the authorities were struggling to address through road building and more parking.

The trigger for change, according to several Freiburgers I met, was a plan by the Federal government to build a nuclear power station 20 miles (30km) away. The successful campaign against it united radical students, church leaders and conservative farmers, giving a boost to the nascent green movement in Germany and changing public attitudes towards a range of environmental issues.

In 1969, Freiburg city council published a transport plan: this followed the traditional approach to providing for the car but also recommended expansion of the tram network at a time when most European cities were dismantling theirs. The closure of old lines was halted and in 1978 work began on a new extension, designed with greater priority over the traffic – a process which has been continuing ever since.

In 1973, the historic city centre, the Altstadt, was pedestrianized (Figure 12.1). This area about half a mile square was one of the largest retro-pedestrianization schemes at the time. Like many similar exercises it was opposed by the business community, which was placated to some extent by a ring of new underground and multi-storey car parks around the edge of the Altstadt.

Construction of the first cycle paths began in 1972. In 1979, a new transport plan was approved, giving priority to walking, cycling and public transport. This was followed by a land use plan that aimed to concentrate new development around public transport stops. These principles have been maintained and refined over the following decades. In 2002, Freiburg became the first German city to elect a Green mayor, Dieter Salomon, who remained in power over a decade later, and the chief planner Wulf Daseking had been in post since 1984, giving the city a degree of stability in its planning and transport policies, supported by a weight of public opinion which often pushed the politicians and planners further.

The council's transport policy is based on five pillars: the emphasis and choice of names has changed slightly over recent years but the underlying principles have remained broadly similar:

Figure 12.1 Münsterplatz in Freiburg's Altstadt (© Stadt Freiburg)

1. extending public transport;
2. traffic restraint;
3. promoting cycling;
4. promoting walking;
5. liveable streets;

and we look at the first three of these in some detail in the following sections.

1. Freiburg: extending public transport

During the first half of the 1980s, there was little sign of a public transport renaissance in Freiburg. Car ownership was rising rapidly like everywhere else. Patronage of public transport was static and costs were rising. The first of the new-style tramlines opened in 1983 and in 1984 the city's public transport operator introduced a low-cost monthly season ticket (the 'environmental card'). This covered trams, buses and trains at all times and was transferable among family members. It was probably the most important factor behind the dramatic rise in public transport use shown in Figure 12.2. In 1991, the season ticket was extended to two adjoining districts (making an area 40 per cent bigger than Greater London). It remained competitively priced at €53.50 a month (about £43) in 2014.

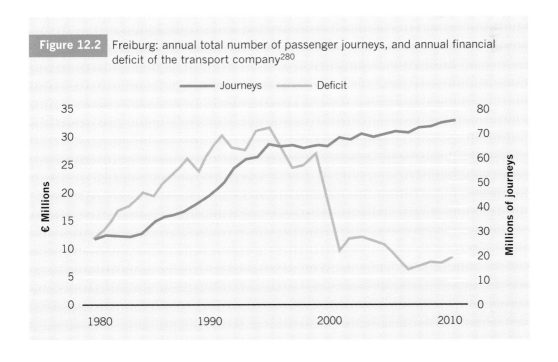

Figure 12.2 Freiburg: annual total number of passenger journeys, and annual financial deficit of the transport company[280]

Unlike in British cities, a single municipally owned public transport operator (VAG) still controls all public transport in the city. This simplified the introduction of the season ticket, although VAG initially opposed its introduction, fearing the financial consequences. Its resistance was overcome by financial guarantees, also offered to the smaller operators in the surrounding areas: where the city agreed to cover any losses incurred for a transitional period. As illustrated in Figure 12.2, the deficit continued to rise until the mid-1990s, since when it has fallen to comparatively low levels. The figure for 2011 represented 13 per cent of operating expenses, excluding separate reimbursements for concessionary fares and school transport. Like-for-like comparisons are difficult to obtain, but this seems to compare favourably with the cost of supporting public transport in British cities.

The tram network forms the backbone of the public transport network, carrying around 70 per cent of passengers. The combination of line extensions and development around the tramlines meant that by 2013 about 80 per cent of the city's population lived within 800 m (½ mile) of a tram stop. The extensions shown in Figure 12.3 aim to raise that proportion to 83 per cent of residents, and also 88 per cent of jobs, within 600 m of a stop.

The density of tramlines within a compact city is an important difference in comparisons to the Manchester Metrolink (Chapter 5), which provides a longer-distance ser-

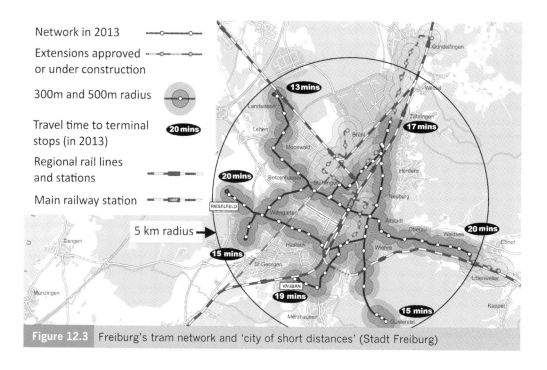

Figure 12.3 Freiburg's tram network and 'city of short distances' (Stadt Freiburg)

Figure 12.4 Regional rail line serving Breisach and the Kaiserstuhl wine-growing region

vice over a wider area. Small towns and villages are linked to Freiburg by buses and the S-Bahn rail network (Figure 12.4), which has improved in recent years. Half-hourly frequencies were the norm in 2014, with bigger improvements under discussion as this book was written.

The differing levels of local authority control in British and German cities has another consequence. In Manchester privately operated buses compete with trams along the same routes. In Freiburg the buses feed the tram system, filling in the gaps and serving outlying areas. Only trams are allowed to cross the Altstadt, sharing space with pedestrians and cyclists along the main shopping street (Figure 12.5).

Figure 12.5 Trams and pedestrians mix on Freiburg's main shopping street

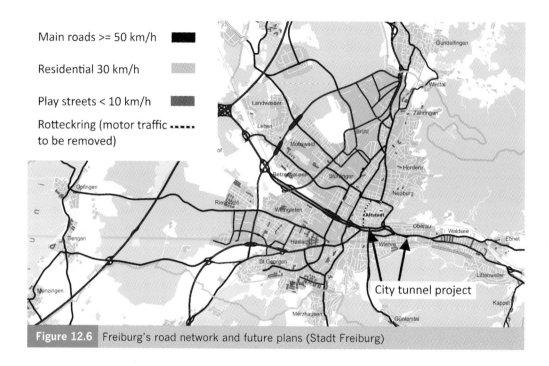

Figure 12.6 Freiburg's road network and future plans (Stadt Freiburg)

The routing of the trams through the city centre has been controversial in the past. A referendum in 1999 was inconclusive. There is a trade-off between direct access to the heart of the city, pedestrian priority and travel times. The volume of pedestrians along the main shopping street slows the trams. From where I was staying in the south of the city, it was much quicker to cycle to the centre than take the tram, despite frequencies of five to seven minutes. One of the extensions shown in Figure 12.3 will help some of the routes avoid this bottleneck. The traffic-free central area will be extended westwards to Rotteckring (see Figure 12.6), creating a more rapid tram route along the western edge of the Altstadt.

2. Freiburg: traffic restraint

The removal of traffic from the Altstadt is one of several forms of traffic restraint in Freiburg. The relatively limited network of roads illustrated in Figure 12.6 carries the through traffic, mainly at speeds of 50 km/h (31 mph). Most of the city's streets, housing over 90 per cent of its population, are limited to 30 km/h (18 mph) or less. There is also a growing number of 'play streets', where pedestrians have priority and vehicles may only move at walking pace.

There are longer-term plans to bury the main east–west road in a tunnel, further removing traffic from the central area but at the same time making through journeys more rapid and convenient. The idea does not seem to have progressed very rapidly since my visit in 2006: cost is the main constraint. The aspiration to remove urban traffic but facilitate inter-urban movement by car is consistent with national practice in Germany. The junction to the west of the city connects to the motorway network, much of which has no speed limit.

Parking is controlled across the whole city, with residents' parking schemes in most residential areas. On-street parking has been removed from the Altstadt, and elsewhere is charged at three rates:

Inner zone:	€2.50 per hour
Intermediate zone:	€1.80 per hour
Rest of the city:	€0.80 per hour

These rates (from just over £2 to about 70p per hour) are considered expensive by German standards. Charges in British cities are often higher although the controlled areas are rarely so extensive, outside London. Rates in Southampton ranged up to £2.50 an hour in 2013, for example, but outside the centre it is still possible to park on the street free of charge.

3. Freiburg: promoting cycling

Like several of the cities discussed in Chapter 7, Freiburg has been developing its network of cycle routes since the 1970s with a combination of on-street cycle lanes, shared paths, junction priority measures, cycle bridges (Figure 12.8), traffic-calmed streets and contraflows along one-way streets. The network (Figure 12.7) is considerably more permeable than the one for motorized through traffic shown in Figure 12.6.

There is nothing unusual about any of these features: all of them can be found in British cities. The key differences are:

- higher and more consistent design standards (fewer botched compromises);
- greater priority over general traffic, particularly coming from side roads and entrances (Figure 12.9), the fact that all the routes join up in all directions.

All of this took time. Three decades of incremental improvements had created the network which astonished me when I arrived in 2006, and the process has continued since then. A €8.9m programme of improvements was planned up to 2020, with the largest element (€2.6m) allocated to a new separate cycle path running alongside a railway line, providing a direct link into the city from the west.[281] Cycle routes are now classified into three categories, shown on Figure 12.7. The priority routes act like main roads for bikes: they are direct, high volume and have minimal interrup-

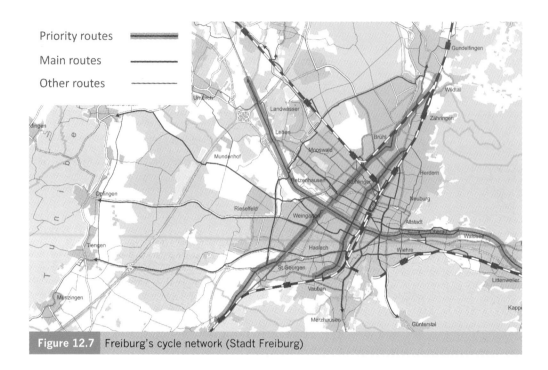

Figure 12.7 Freiburg's cycle network (Stadt Freiburg)

tions. To deal with the width constraints discussed in Chapter 7, some streets have been made one-way for general traffic, freeing space for wide cycle lanes in both directions and replacing pavement cycle paths to give more space to pedestrians. As in many European cycling cities, Freiburg's policy has changed from creating pavement paths to replacing them, because of the conflict they cause between cyclists and pedestrians.

While car parking has been constrained, cycle parking has expanded – around tram stops, neighbourhood centres and the city centre. There is a particularly impressive cycle parking station with a cafe and repair shop by the main railway station (Figure 12.10).

The 5 pillars of Freiburg's transport policy are complemented by 12 principles of urban planning,[282] of which the 3 most relevant to transport are:
1. a city of short distances;
2. high-density development along public transport routes;
3. a city of neighbourhoods.

The city of short distances can be seen in Figure 12.3. Despite being surrounded by mountains and forests on two sides, the vast majority of the city's population live within a 5-km (3-mile) radius, with outlying suburbs clustered around the tram

CHAPTER 12 European cities: inspiration and similarities

Figure 12.8 Cycling bridge formerly open to car traffic

Figure 12.9 Pavement paths have priority over side entrances and roads

lines. The city's land use plan favours major development around the city centre and smaller shops in neighbourhood centres. Car-based retail developments on the outer fringes are prohibited.

With population rising in the early 1990s, work began on two new suburbs, Vauban and Rieselfeld (see Figure 12.7). Vauban, the smaller of the two, is one of the carfree developments described in the next chapter. Rieselfeld, built on a former sewage works, now houses around 10,000 people, mainly in apartment blocks. As the city owned the land, it was able to control the development process, selling small parcels of land to builders and *Baugruppen*, groups who collectively build homes for

Figure 12.10 Cycle parking and facilities by Freiburg's main railway station

Figure 12.11 Rieselfeld: main street

Figure 12.12 Residents' parking spaces

themselves. The tramline that follows the district's main street began running at an early stage, when there were only approximately 1,000 people living there (Figure 12.11). Crucially, and unlike comparable suburban developments in Britain, parking is controlled across the whole district. On-street parking is limited and charged at the outer-city rate. Most residential parking is underground or in covered parking areas, with most dwellings allocated a single space (Figure 12.12). As a result, car ownership at 283 per 1,000 people was less than the average for Freiburg and just over half the national average.

Freiburg – conclusions

So what were the effects of all these policies? Figure 12.13 shows the changes in modal share up until 1999, the last time the city conducted a city-wide survey. These figures show the travel by residents within the city but their pattern of total travel was very similar; the city is fairly self-contained, with residents making relatively few trips to other places.

Clearly the policies achieved their main aim, to change the way people travelled. However, this graph does not tell the whole story: if we look at the number of trips, instead of proportions, cycling and public transport both increased while car journeys remained stable, despite a rising population. Walking as a main mode fell, however, as people cycled or took the tram – much easier with a season ticket which costs nothing each time you board. Overall, Freiburgers now travel more often: the total number of journeys per person rose by around a quarter from 1979 to 1999.[283] All of this is consistent with the discussion in Chapter 5: improving public transport has several effects and encouraging people to travel more often is one of them.

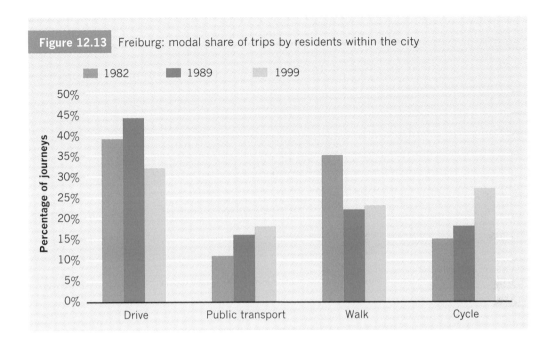

Figure 12.13 Freiburg: modal share of trips by residents within the city

Car ownership was rising strongly in the 1980s and has stabilized more recently at around 390 per 1,000 residents, much lower than the national averages for Germany and Britain, although higher than some less affluent British cities. The pattern of car ownership has been changing, rising among older people and falling among people under the age of 50. Under the age of 25, car ownership has fallen to negligible levels. Students, who make up nearly one in seven of the city's population, made over half of their trips by bicycle in 1999.

Figure 12.14 Market on Freiburg's Münsterplatz

Figure 12.15 Münsterplatz, in Freiburg's Alstadt in the 1960s (© Stadt Freiburg)

Just as important as all that, Freiburg feels good to visit and move around, particularly by bike and tram. There is still some congestion at peak times, but conditions for drivers certainly don't seem any worse than in British cities of a similar size. Walking around the Altstadt, it is difficult to imagine it was once a giant car park (Figure 12.15). With traffic running through its streets and parking all over the Münsterplatz, you wouldn't be able to fit all the people who move around it today (Figure 12.14). This is generally true of such changes: whatever the arguments at the time, few people would ever want to go back to how it was.

Groningen, Netherlands

A city which experienced one of the most vituperative disputes about traffic changes in its city centre was Groningen, in the north of the Netherlands. Its population is 195,000, making it about the same size as York, slightly smaller than Freiburg. Though not very well known outside the Netherlands, it has been nicknamed the 'bicycle capital of the world'.[284] Among the "major Dutch cities" it has registered in most years the highest modal share of cycling.[285] I thought Freiburg's cycle infrastructure was likely to be the best in the world, until I toured the Netherlands and stayed in Groningen the following year.[286]

During the 1960s, Groningen was a city with rising car ownership – above the Dutch national average. As in British cities, the authorities responded with plans to demolish many buildings to make room for new highways. Then in the early 1970s, a coalition of leftwing parties won control of the city council with a programme aiming to prioritize pedestrian movement and remove through traffic from the city centre. In 1977, the centre was divided into four sectors. Motor vehicles could drive in and out of each sector but not between them. A ring of old roads more or less followed the circle of canals that surround the city centre. Further out, the new highways created a partial inner ring road that was never completed, though with existing roads they formed a ring of sorts, which through traffic could follow (see Figure 12.20).

As in many British cities, these plans were strongly opposed by many businesses. A group of 80 business owners in the city centre appealed, unsuccessfully, to ministers and the King to get the changes reversed. Two months later, a small business association published a survey showing a fall of 30 per cent in sales in the two months after the changes. If this was accurate, many of them would have rapidly gone out of business. This was the first of several business surveys, including one conducted by a university for the city council in which a quarter of respondents reported a fall in turnover. All of these surveys were conducted after the event and were explicitly related to the traffic plan. This made them vulnerable to two common research problems: self-selection bias, where those with an axe to grind are more likely to respond; and policy response bias, where people answer in a certain way hoping to influence public policy. To reduce the risk of such biases, researchers usually try to present surveys in a neutral way, so they are not perceived in relation to a policy controversy. In Groningen, the regular business survey conducted by the Chamber of Commerce in 1978 provided a point of comparison: it found profits had increased faster in the city centre than elsewhere in the province. Whatever the truth about the effects on trade, many businesses continued to lobby unsuccessfully for several years to reverse or water down the scheme.

Residents' groups appeared more satisfied with the changes and the council took other measures to help local businesses, like reintroducing street markets, which had been forced out of the centre by traffic and parking. A few businesses threatened to move out of the centre; a later survey showed that only four had done so. In a pattern typical of many other cities, businesses handling larger items, such as wholesalers, gradually moved out of the centre and were replaced by shops and service-sector businesses.

The feeling that the council wasn't listening soured relations with the business community, which took time to heal. But heal they did, and gradually the Groningen Chamber of Commerce became one of the greatest supporters of traffic-free city centres. So much so that when I visited in 2007, it was sponsoring a network of free guarded bicycle parks in the centre. The bicycle is now used by a third of its customers, compared to one in five who drive, mainly from outlying areas.[287]

A few years later, when doing some research to inform a similar debate in Bristol, I had the following email exchange with the Groningen Chamber of Commerce:[288]

> **Extract from email exchange with Groningen Chamber of Commerce**
>
> What is the attitude of the chamber and its members towards traffic management in the city centre today?
>
> "Nowadays our members and the entrepreneurs in the city center are happy that the traffic is banned. The number of visitors has increased, the centre of Groningen is the number one shopping centre of the Northern Netherlands and Groningen is the city of cycling. The number of terraces in summer has increased and the city of Groningen is the place to be."
>
> Do any businesses today advocate reopening the city centre to through traffic?
>
> "No one. It is no issue!"

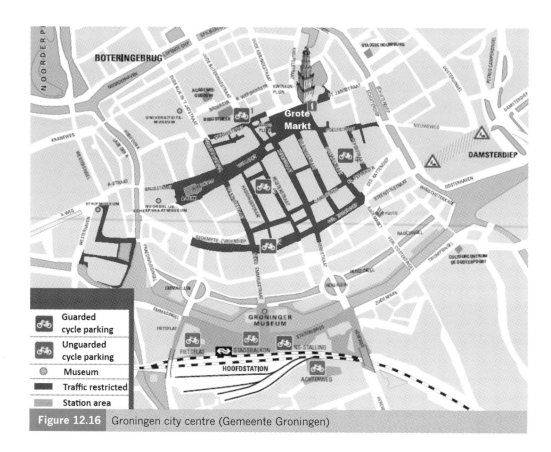

Figure 12.16 Groningen city centre (Gemeente Groningen)

Today in the city centre many of the streets, including the central square (Figure 12.17) are fully pedestrianized, and a range of time restrictions on vehicular access applies to others. There are 11 multi-storey car parks in and around the centre, but very little on-street parking. The controlled parking zone has been gradually extended, though not as far as the outer suburbs. Within the city centre, the vast majority of moving vehicles are bicycles and the picture painted by the Chamber of Commerce is no exaggeration: removing motor traffic has made the centre of Groningen a great place to visit and spend time. There is no need to close roads for street events, like the one shown on Figure 12.18, as plenty of traffic-free space is always available.

The population of the city centre has been rising rapidly in recent years, reaching 18,000 by 2013. Many of these people live on streets where vehicle access is limited. There are some residents' parking spaces, but less than a third of households in the centre own a car anyway.

CHAPTER 12 European cities: inspiration and similarities

Figure 12.17 Groningen Grote Markt – Central Market Square

Figure 12.18 Street festival on Groningen's Grote Markt

As in Freiburg, transport policy has favoured cycling, walking and traffic restraint (Figure 12.16), supported by planning policy to maintain a compact city with neighbourhood centres. There is less emphasis on public transport, however: there is no tram system and buses make only a modest contribution.

The quality and comprehensiveness of the cycle network are awe-inspiring. Key principles are:
- directness;
- priority;
- consistent widths;
- three-way separation of cars, bikes and pedestrians.

It was in Groningen that I first became aware of filtered permeability as a transport planning principle. Cor van de Klaauw, the senior transport planner at the time, described Groningen's approach as creating a 'fine grain' for cyclists and a 'coarse grain' for motor vehicles, as you can see by comparing Figure 12.19 and Figure 12.20.

The suburb of Beijum where I stayed for several days illustrated the approach. There are only two ways in and out for general traffic, both of which come out on the same road, with an orbital road pattern making the route from Beijum to the city centre circuitous. At the same time, there are many short-cuts for cyclists and pedestrians and what might almost be described as a cycling motorway leading directly to the city centre. These 'priority routes' mean that cyclists from most of the newer suburbs can travel to the city centre without having to mix with traffic or stop at traffic lights. Most junctions are designed to provide some form of priority; the forms vary according to the situation. Where side roads meet main roads, pavements are sometimes continued across the junctions as 'raised tables' (Figure 12.21). Where cycle routes

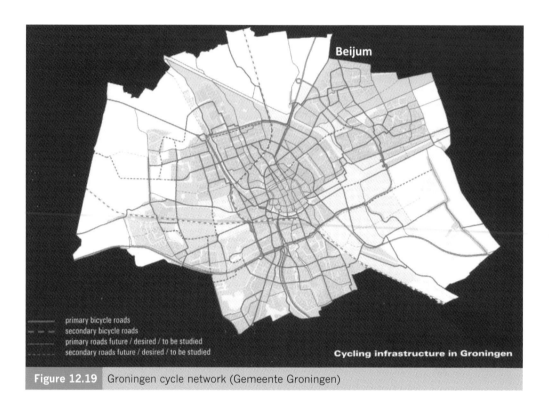

Figure 12.19 Groningen cycle network (Gemeente Groningen)

Figure 12.20 Groningen network for through traffic (Gemeente Groningen)

cross some busier roads, either bridges are used or traffic lights are phased to go green for cyclists only, while all other traffic is held on red (Figure 12.23).

Filtering to give the advantage to the bike is provided by bridges over canals, and by devices like bus gates – where buses and bicycles are the only vehicles allowed through. Off-road cycle paths often have a separate pavement for pedestrians (Figure 12.22). All of this has fostered a culture where cycling is normal for everyone, and is used for a much wider range of purposes than in Britain. People with disabilities also use the cycle network with a range of vehicles including battery-operated mobility scooters and hand-powered recumbent bicycles.

I asked Cor van der Klaauw how they had achieved all this. It helped that all his staff and most of the councillors cycled to work, he explained: he worked with people who understood the issues. Though some transport issues in Groningen have been divisive, a broad consensus has always supported the principle of promoting cycling.

One of the problems this has created is cycle parking. A new 4,100 space cycle park by the main railway station (Figure 12.24) had just been completed when I arrived, but by the time it opened, it was already insufficient. A temporary 'bicycle tower' holding a further 2,000 has become a permanent feature and the council has continued to look for opportunities to increase cycle parking since then.

Between 1964 and 1980, motor traffic crossing the inner ring-road cordon fell by nearly a half, although traffic continued to rise on the outer cordon as the city grew in size. Historical figures for modal share are not available, so it is difficult to assess what effect the transport policies had on overall travel. The results of the latest surveys of the city and the province are shown in Figure 12.25.

Figure 12.21 'Raised table' junction

Figure 12.22 Cycle path with separate pavement

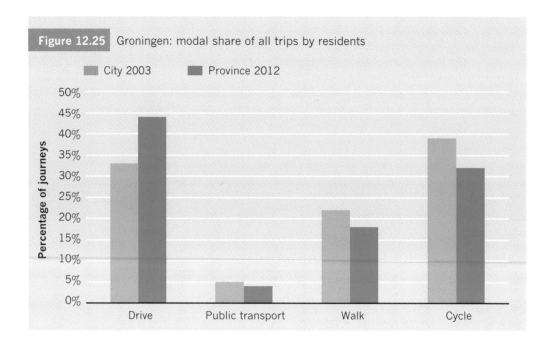

Figure 12.25 Groningen: modal share of all trips by residents

Car ownership (352 per 1,000) is now about three-quarters of the national average: 47 per cent of households across the city do not own a car. As in Freiburg, students form a high proportion of the population (17 per cent[289]) but unlike Freiburg, household income is lower than the national average. The difference in transport trends between the inner city and the outlying areas has continued to widen in recent years: since the last city survey in 2003, traffic counts around the inner cordon have shown more bikes and fewer cars but further out the car traffic is increasing and the bike traffic is reducing.

Figure 12.23 Groningen inner ring road: cycling only traffic light phase

Figure 12.24 Cycle parking at Groningen Railway Station

This combination of a growing city with rising levels of traffic on its outskirts prompted the city, provincial and regional authorities to consider a solution which seemed to have worked well elsewhere: a tram system. A feasibility study was commissioned in 1997 and when I visited 10 years later, the plans appeared to be progressing smoothly, though no decisions had been made about the route. As the plans progressed the project became more controversial. The same issue which dogged the debate in Freiburg – where to run the line through the city centre – provoked the first organized opposition. In an echo of the 1970s, traders along the chosen route launched a campaign against the trams, only this time the customers they feared they would lose were pedestrians and cyclists on streets already closed to cars. In 2012, with the council's finances under pressure, the project brought the ruling coalition down when two of its parties refused to sanction a budget increase, much to the annoyance of the province representing the outlying towns. The province and city had reportedly spent €40m by the time the project was scrapped.[290]

All of this may sound rather familiar to people in several British cities. In west London an ill-fated tram proposal caused a change of leadership in the borough of Ealing. In Bristol two local authorities couldn't agree on the route before the government changed the funding rules, effectively closing the door on new tram systems.

Lyon, France

One country that has invested heavily in tram and light rail systems in recent years is France. Over 20 cities have built new systems since the mid-1980s. All the conurbations of over 500,000 people, and several much smaller cities (eg Orleans, population 113,000), now have a tram or metro system.

Lyon, a city I have visited several times with postgraduate students, provides an interesting contrast to Freiburg and Groningen. Its leaders have pursued similar aims to change travel behaviour and reclaim city centres from motor traffic. They have used some similar means but with more emphasis on public transport investment.

Greater Lyon is the third largest conurbation in France, with a population of around 1.3 million – similar to Merseyside. By the end of the 20th century, Lyon had four underground metro lines (built between 1974 and 1991), trolleybuses (Figure 12.26), and two historic funicular railways.[291] Since 2001 two of the trolleybus lines have been partially segregated from general traffic and four new tramlines have been built. In some places short tunnels have been dug, or old rail lines reused; elsewhere the tramlines have taken lanes away from general traffic (Figure 12.27).

Although trams and trolleybuses are less expensive than underground rail, the cost to local taxpayers has been considerable. The capital budget of €1.1bn from 2009 to

Figure 12.26 Lyon: trolleybus

Figure 12.27 Tramline reducing traffic to one lane

2014 was just under €1,000 per inhabitant of Greater Lyon. Unlike similar projects in Britain, central government provides only a tiny share. Figure 12.28 shows the combined capital and revenue budget for Sytral, Lyon's public transport operator.

The largest element, the transport tax, is levied on employers as a percentage of salaries. This tax has been running since the 1970s, providing the authorities with a fairly predictable income stream against which they can also borrow. Needless to say, the cost of servicing this debt has been growing along with the network: 18 per cent of the budget for 2013 went on loan repayments.

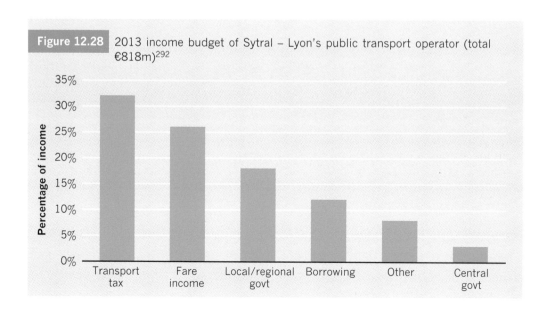

Figure 12.28 2013 income budget of Sytral – Lyon's public transport operator (total €818m)[292]

CHAPTER 12 European cities: inspiration and similarities

Figure 12.29 Lyon: hire bikes

Figure 12.30 Cycle paths along the river Rhône

Not all of the transport changes have cost large sums of money, and one of them came at no cost to the taxpayer. In 1998, Rennes became the first French city with an on-street bike hire system, and Lyon followed suit in 2005. The bikes (Figure 12.29) are provided and maintained by an advertising company, JCDecaux, which was awarded a franchise giving it sole access to the bus stop advertising space and other advertising boards owned by the city. In the first 2 years, 60,000 people signed up, using the bikes for an average of 20,000 movements a day, mainly for commuting.[293] The bikes are well used in the central areas (though they are not always very well maintained), but an important effect was on the visibility of cycling, which had been a marginal activity in Lyon before then. Rates of cycling nearly doubled in the two years after the system started, mainly because more people started using their own bikes.

The city has also made some attempts to improve its cycle network, though not as seriously as Groningen or Freiburg. Conditions are similar to many British cities, with two high-quality joined-up routes following a river (Figure 12.30) and a tram-line (Figure 12.31), and many fragments of varying quality, considered fair game for illegal parking (Figure 12.32).

As in most French cities, the investments in public transport were accompanied by improvements to public open space. The most spectacular change was transformation of the banks of the river Rhône from a car park into a 5-km (3-mile) long linear park with the cycle and pedestrian route shown in Figure 12.30. After many years of discussion, work began in 2003. To compensate for the loss of parking spaces an increase in underground parking was included, but, as you won't be surprised to hear, opposition grew as the details became clearer. Parking on the riverbanks had been free; the new underground car parks would not be (the central car parks cost €2.10 an hour in 2013). Shopkeepers interviewed by the local press cried "no cars, no

Figure 12.31 Tramline and cycle path near Lyon Part-Dieu Station

Figure 12.32 A blocked cycle path in Lyon

customers – it will kill the shops in the city centre".[294] An opposition councillor organized a survey: most of the people responding opposed the loss of riverside parking.

The cost of the project began to rise, partly because of changes made to placate objectors and also because services like a fibre optic cable had to be moved. Opposition grew among the various authorities which made up Greater Lyon. Gérard Collomb, Mayor of Lyon, wrapped up one debate with the comment:

> "If my fellow mayors want the project to stay in its draft state, I will listen to them, but come the day of the inauguration, they will all be jostling for a place in the photo."

Despite the opposition, the transformation went ahead (Figure 12.33). Online videos show the mayor at the long-awaited inauguration in 2007, flanked by the great and the good, expounding in a style reminiscent of Général de Gaulle.[295] Images reminding voters of how much the riverbank had improved (Figure 12.34) were used in the mayor's re-election campaign, contributing to a victory with three times the vote of his nearest rival. The perceived success of the strategy encouraged the city's leaders to press ahead with a similar scheme on the banks of the city's other river, the Saône, opening in 2013. Although some riverside residents complain of late-night revellers and young people on skateboards, even they would not want to go back to a car park along the banks.[296]

National planning policy in France has, for many years, sought to rein in urban sprawl. Complex layers of regional and city-region authorities have been set up and charged with powers to put these national policies (which most of them agree with) into practice. But although French cities have succeeded in regenerating and intensifying their

CHAPTER 12 European cities: inspiration and similarities

Figure 12.33 Banks of the Rhône in Lyon

Figure 12.34 A municipal poster reminds voters how it used to be

city centres, sprawl in the surrounding areas has continued unabated. Apart from the usual market pressures and lobbying from developers,[297] local government in France is fragmented into thousands of unusually small *communes*, each with a mayor who has the final say on what gets built where. The city councils control only a relatively small part of each city. And unlike Britain, where suburban and rural local authorities resist new development, the mayors of small communes in France want more of it. Although Greater Lyon is often considered one of the stronger metropolitan authorities, and the population of the city has been increasing in recent years, the small communes in the rest of the region have been growing even faster.[298]

All of this has produced a divergent effect on travel behaviour, similar to the one observed in Groningen. The last city-wide survey of travel patterns was conducted in 2006. Most, though not all, of the tramlines were operating at that point, and the parking had been removed from the banks of the Rhône, though the cycle and walking route was yet to open. Though less dramatic than in Freiburg or Groningen, the share of car journeys, shown in Figure 12.35, did fall within the conurbation of Greater Lyon. Given the scale of investment, the increase in journeys taken by public transport was relatively modest. Walking and cycling together made a greater contribution to the fall in driving, although the proportion of journeys taken by bicycle alone remained very low. At 460 cars per 1,000 people, car ownership was higher than Freiburg or Groningen, though slightly below the national average for France.[299] For Lyon and Villeurbanne, the two most urban communes, the figure was 420 per 1,000. This level remained unchanged between 1995 and 2006, when car ownership across the country was rising.

The 2006 survey (Figure 12.35) also covered a wider sub-region around Lyon. In the surrounding towns and rural areas it tells a rather different story. Although the

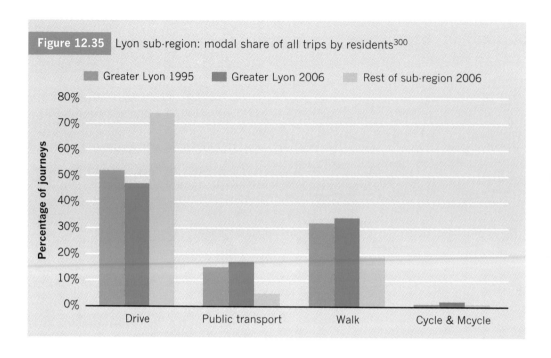

Figure 12.35 Lyon sub-region: modal share of all trips by residents[300]

rail network carries a majority of commuters from some towns such as Mâcon and Belleville, overall most people outside the main urban area depend very heavily on their cars. Like most French cities, Lyon is surrounded by motorways and whereas car use is falling in the larger cities, it is rising elsewhere.[301]

Freiburg, Groningen and Lyon compared

If we now compare the three cities described in this chapter, connectivity and integration between walking, cycling and public transport are key similarities, which I have also observed in many other European cities. Tickets valid across all modes of transport, cycle parking at stations and tram stops, and buses that feed tram networks are just a few of the many examples of this. They are all made easier where a single authority controls all the elements; where city authorities have to negotiate with their neighbours or business interests, integration becomes much more difficult and public money may be wasted on aborted projects or ineffective compromises. In none of the three countries do business interests have the formal decision-making role that they have in Britain – on the boards of local transport bodies, for example. Businesses may lobby but they don't get a vote on how public money is spent. Cost-benefit analysis is sometimes used to compare alternatives but this is given less weight than in Britain; long-term strategies are more important.

A combination of measures was key to the success of all three cities. If just one measure is taken, public transport, cycling or walking may substitute for each other instead of reducing car driving. Or people may simply choose to travel more. This can be avoided if all three are supported at the same time and car driving is restrained in some way. That said, it is possible to focus more on cycling, as in Groningen; more on public transport, as in Lyon; or equally, as in Freiburg. Chapter 13 will provide an example of a smaller town where walking has been the main focus, Louvain-la-Neuve in Belgium.

Public transport is a lot more expensive to promote than cycling or walking. Before too long, cities like Lyon may have to rein in their capital investment programmes and start repaying some of the debts they are accumulating. Similarly, the Groningen model is most appropriate for a smaller city with a favourable terrain. Groningen has managed without any big investment in public transport so far, but if it continues to expand, something like the tram system will inevitably rear its head again.

Although all of the cities planned for, and achieved, a shift towards sustainable modes of transport, unintended consequences were often at work. The fall in walking (as a main mode) in Freiburg was a negative consequence. The faster growth of walking compared to the use of public transport in Lyon was a more positive example.

All three cities have restrained motor traffic in different ways, Groningen and Freiburg more so than Lyon, which may explain why the scale of modal shift has been less dramatic in Lyon. Proposals for restraint always provoke opposition, from business interests and from individuals, particularly where parking is affected. Though the public in Freiburg and Groningen have been more supportive, the leaders of the cities have had to stand firm against concerted opposition at times. Opposition is not always direct. In some cases, a reasonable-sounding compromise may undermine the purpose of a change. For example, when opponents of the changes in Groningen realized the council was determined to go ahead with the four-sector plan in 1977, a group of businesses came up with a compromise, which would have opened a breach between two of the sectors, undermining the purpose of the plan.

None of cities has used congestion charging or any other financial instruments to discourage car use, apart from parking charges, which are generally lower than equivalents in British cities. The changes in all three cities produced improvements clearly visible to everyone. And whatever the transport or financial arguments, trams seem to integrate more easily than diesel buses into pedestrian-friendly streets. All of this helped to promote a supportive consensus after the changes, if not before.

The unintended consequences mentioned above raise two obvious questions: "How much difference did the policies make?" "Were any other factors involved?" The answer to the first question is "We can't be sure" and to the second "Almost certainly". Comparable information is not available to explore all the possible explanations fully. (This will be easier for the British cities described in Chapters 14 and 15.)

Some comparisons can be made of demographic profiles. The trends in the three cities have been quite different in recent years. In Groningen, the population became much younger, with an increase of over 30,000 in the 20 to 29 age group between 1995 and 2009.[302] The population of Lyon, on the other hand, became slightly older between 1995 and 2006.[303] In Freiburg, the proportion of 18-25 year olds has grown fractionally in more recent years, whereas the 25-40 age group declined slightly.[304] More importantly, car ownership in both these age groups declined between 1995 and 2011.[305]

The perceptions of Freiburg and Groningen as 'university cities' is sometimes used as an excusatory myth – a reason why their policies 'wouldn't work here'. Most cities have some form of educational institution and the number of students is generally too small to make much direct difference to city-wide travel. As we will see in the analysis of Cambridge in Chapter 15, universities *can* help to influence the transport cultures of the cities where they are based, although there is nothing automatic about that process.

Conclusions

After each of my European summer tours I returned with strongly mixed feelings. Inspiration was one of them. In each of these cities and several others (Basel, Bordeaux, Amsterdam, Maastricht, Münster, Cologne, Malmö, Odense, Copenhagen and Louvain-la-Neuve) I saw policies which worked better than they do in comparable places in the UK. As a visitor I felt privileged to cycle along rivers, stroll through squares and drink coffee on terraces local people had either fought to liberate or to preserve as stinking car parks in the past. The pleasure of my first English real ale after months away was swiftly followed by frustration. "There's no rocket science in any of this," I would say to anyone willing to listen, "I've seen nothing that couldn't be done here." This comment would also apply to the carfree developments described in the next chapter.

CHAPTER 13
Carfree developments

Louvain-la-Neuve, Belgium – The 'pedestrian town'

Cycling through Belgium in the late 1980s, I stopped at a cafe near the Dutch border and asked the waitress if she spoke French. "*Un peu*," she replied in a voice which suggested I'd better not ask for anything complicated. A gruff voice behind me barked, in French, "You a *francophone*?" I turned round to face a rather large man, looking me straight in the eyes. "No," I replied (still in French), "I'm English but I'm living in France at the moment…" At this, his expression brightened and breaking into English he said, "Oh, you're English – that's much better. Come and have a drink with us."

I may have been over-influenced by things I've read and particular people I've met, but in later visits to Belgium I have noticed the gulf growing wider between its two linguistic communities – Walloon (French) and Flemish (Dutch). The superior cycling infrastructure in Flanders is partly due to the Dutch influence and partly to a growing economic advantage over their French-speaking neighbours. But one of the most radical experiments I have witnessed emerged on the French-speaking side after one particularly sorry episode in this history.

In 1967 conflict between the two communities came to a head with violent demonstrations by Flemish students and nationalist parties demanding the expulsion of the French-speaking section of the bilingual university of Leuven.[306] This *affaire de Louvain*, which brought down the government of the day, was resolved by an agreement to split the university with a new campus in French-speaking Wallonia.

The trauma unleashed "a new spirit of utopianism: a rupture to prepare for a better future".[307] The French-speaking vice-chancellor of the bilingual university, Michel Woitrin, was put in charge of the project. Woitrin was partly influenced by the American campus model, but also by 'town and gown' interactions in Oxford, Cambridge, and the old town of Leuven. A new campus was not enough. His ambition was to build the first and only new town in the history of the country. A site was chosen 20 miles (32 km) southwest of Brussels for Louvain-la-Neuve (LLN – 'New Leuven').

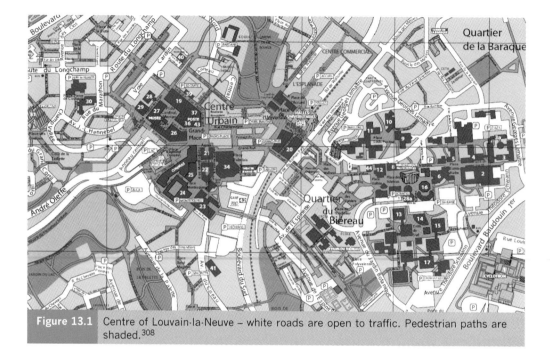

Figure 13.1 Centre of Louvain-la-Neuve – white roads are open to traffic. Pedestrian paths are shaded.[308]

In common with the British towns built under the New Towns Act, the state bought or compulsorily purchased the land at 'existing use' (mainly agricultural) prices. The Flemish authorities, eager to bring an end to the crisis, helped with the funding.

Another influence on Woitrin was Buchanan. The masterplan for LLN was arguably the purest example of Buchanan's concepts of traffic separation and environmental areas (discussed in Chapter 11). In LLN the pedestrian is king. The most direct routes across the town and particularly in the centre are reserved for pedestrians and cyclists only (Figure 13.1). Cars and other traffic have to fit round the outside or underneath the concrete plinth which bears the town centre.

Woitrin also wanted to separate cyclists from pedestrians but failed to persuade the triumvirate of colleagues who worked with him. There is now a network of cycle routes leading into the town, but within the central areas cyclists and pedestrians mix.

The first inhabitants arrived in 1972. Since then, LLN has grown, more slowly than originally envisaged, but steadily, to around 20,000 in 2013, of which roughly half are students.[309] Of the rest, only a minority now work at the university. There are employment areas and a medium-sized shopping centre. It has become, in other words, more like a normal town. With its lakes and green spaces, it shares some of the better aspects of the English post-Second World War New Towns, but with

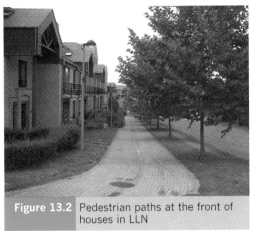

Figure 13.2 Pedestrian paths at the front of houses in LLN

Figure 13.3 Rear car parking

less concrete brutalism and being far less dominated by traffic (Figure 13.6). Several English New Towns (Harlow, Stevenage, Runcorn) and towns which expanded in the 1970s (like Bracknell, Nailsea) have what is known as the 'Radburn layout', where the houses front on to footpaths and green areas (Figure 13.2), with cars parked at the rear (Figure 13.3). This has fallen out of fashion in Britain, partly because in car-dominated towns, people tend to treat the rear as the real entrance. In LLN, the concept works much better. Comparing Figure 13.1 with the map of central Milton Keynes (Figure 13.4) illustrates one of the reasons for this: in LLN the roads fit round the pedestrian paths but in Milton Keynes it is the other way round.

Another important difference is the density of development. Conventional wisdom at the time (particularly in Britain and North America[310]) associated high density with slums and social problems. As a result, the English New Towns were planned to remove people from the major cities, and rehouse them at lower densities.[311] Woitrin resisted this fashion, arguing that interactions – between teachers and students, different cultures, old and young – were essential to make the new town work, and this called for higher densities, to be provided by flats rather than houses. His colleagues were less convinced. The result was a compromise, but one which works much better than the equivalents built in Britain at the time. There is a mixture of housing types including three-storey town houses and apartment blocks between the commercial and university buildings in the centre, arranged around green spaces and footpaths but also squares of different sizes (Figure 13.5), allowing for a thriving outdoor life.

The design team originally envisaged an elevated metro linking the new town with the local hub station at nearby Ottignies. In the end, it proved easier to build a spur off the Brussels to Namur line leading to a conventional station. The railway line (and one of the main roads) runs underneath the town centre. The station is acces-

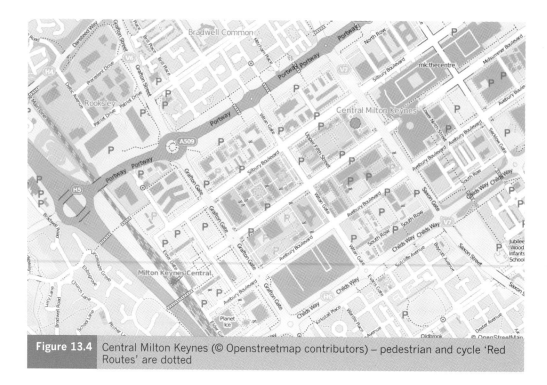

Figure 13.4 Central Milton Keynes (© Openstreetmap contributors) – pedestrian and cycle 'Red Routes' are dotted

sible from the Grande Place at the centre of the town. The bus station is not far away although the connection between the two is not particularly convenient. Plans have been slowly progressing for a regional rail network, like the RER in Paris or the German S-Bahns, but the project has been delayed by financial constraints and the climate of suspicion between Flemish and French-speaking politicians.[312]

Parking is controlled across the whole town, although there is no shortage of it. There are large car parks on the lower levels beneath the central plinth, and rear courtyards in the residential areas. Residents are entitled to two free parking permits and additional permits for €60 a year.

Like the cities discussed in Chapter 12, LLN is directly connected to the motorway network. This dual approach of favouring the pedestrian inside the town and the motorist beyond its limits is reflected in the travel behaviour of its residents. Within the town, walking is the most common mode; driving is much less common (Figure 13.7). External travel is more like the pattern we would expect to find in a small town (Figure 13.8). 83 per cent of the non-student adults had access to a car in 2006; driving is their most common mode of travel although they also make relatively frequent use of public transport. Less than a third of the students have a car, so their patterns are different. The survey of residents from which these two charts are derived asked different ques-

Figure 13.5 Louvain-la-Neuve square

Figure 13.6 LLN route around the lake

tions from those normally used in transport surveys, so the modal shares cannot be directly compared to any of the others shown in this book. However, the pattern is very clear, and obvious to anyone who visits LLN. The 'pedestrian town' is not just a marketing tag: most of the movements within the town are on foot. This is what gives the town its lively, social feel, at least during university term times.

Opinion is divided about LLN as a place to live. Many of the university staff and a few of the students prefer the cosmopolitan big city life of Brussels. Several people I

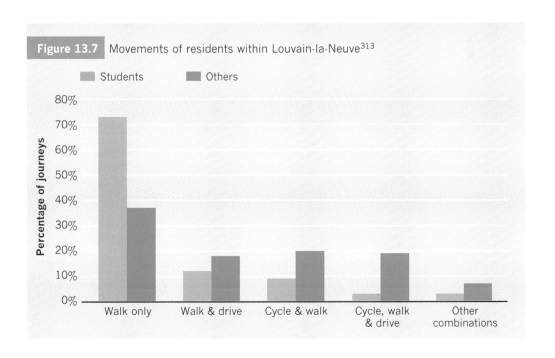
Figure 13.7 Movements of residents within Louvain-la-Neuve[313]

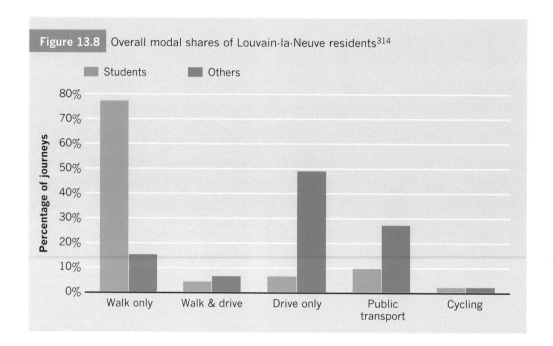

Figure 13.8 Overall modal shares of Louvain-la-Neuve residents[314]

met said LLN felt rather "artificial". In a country with no other new towns, points of comparison were limited. Those who have chosen to live in LLN seem to rather like it. It suits a certain "provincial type", said one of the academics with a little disdain towards some of his colleagues. He preferred to live in a leafy suburb along the train line to Brussels. The trains, he said with a smile "are almost as bad here as they are in Britain", though he still used them to travel to work.

The design of LLN has some drawbacks. It is not the easiest place to find your way around when you first arrive. This was apparently deliberate, and arguably adds to its character.[315] It is a town which has to be discovered. Whether that outweighs the inconvenience for new visitors is debatable, but as a place to stroll around and explore, it certainly beats English New Towns like Skelmersdale or Runcorn. Some academic studies have tried to show that places which are difficult to find your way around discourage people from walking; that may be true in some places, but not in LLN.

For most journeys in LLN, the travel network has been designed to make the route along pedestrian paths shorter than the route by road, but there are a handful of exceptions. Where an occasional stretch of road offers a short-cut, people follow their 'desire lines' along the carriageways. A conventional pavement would be an improvement in those places. Cycling along pedestrian streets like the one in Figure 13.2 feels rather strange to a visitor. Though pedestrians and cyclists seem to mix with no obvious problems, cycling is rather slow and inconvenient – I remember feeling it's a pity Woitrin didn't win that particular argument. There are many paths connecting LLN to surrounding towns but these are even more difficult to follow. The network of major roads around the town has caused all the usual problems, with air pollution emerging at the top of a list of residents' concerns.

For all of those shortcomings, the original intention to create a town for the pedestrian, a place to encourage interactions on the street, has worked well. Changes of fashion (in urban design like many other fields) can blind specialists as much as the general public to positive lessons from the past. It is difficult to imagine any new town or new neighbourhood designed around conventional streets or shared space achieving levels of walking like those in LLN. Michel Woitrin, who died in 2008, could be proud of that.

Carfree developments within cities

Although the car has been pushed to the fringes of LLN, there are no constraints on car ownership there. So would it be possible to go further, to build towns and cities free from cars and motor traffic? There are parts of the world where people have always lived without cars because no road access is possible or none has been provided. Some of these are historic areas or entire settlements, like Venice. In the 1990s, beginning in Germany and Austria another movement with utopian ideals began to promote the idea of carfree new developments. Several of these have been built across various European countries, and were the original reason for my European travels.

Vauban (Freiburg, Germany)

The largest recently built carfree development, Vauban in Freiburg, is home to over 5,000 people. Although it is sometimes called *autofrei* (carfree), the local authority prefers the term *stellplatzfrei*, literally 'free from parking spaces'. Residents have to sign an annual declaration saying whether they own a car or not: car owners have to buy a space in one of the multi-storey car parks around the periphery.

The whole neighbourhood is a large cul-de-sac for vehicles, though pedestrians and cyclists can move in all directions. There are three stops along the tramline in the middle of the main street, which has a 30 km/h (18 mph) speed limit and metered parking (Figure 13.9). Several parking spaces are reserved for Freiburg's car club, which has the greatest concentration of its members in Vauban.

Vehicles are allowed down the *stellplatzfrei* residential streets at walking pace to pick up and deliver but not to park (Figure 13.10), though not everyone abides by these rules. Freiburg is pretty laid back by German standards and you don't see many

Figure 13.9 Vauban, main street

Figure 13.10 Vauban *stellplatzfrei* side street

traffic wardens, so implementing the no-parking rule depends on agreement among residents. That said, you rarely see vehicles moving on the *stellplatzfrei* streets, and in the absence of traffic the kids have taken them over. Roller blading and unicycling were the big crazes when I was there.

I have been to many so-called home zones, which are supposed to provide an environment for children to play in the street, but are really just prettified car parks. A study of another carfree development in Nuremberg confirmed what I observed in Vauban: that children are allowed to play independently at a younger age than they are in more conventional developments.[316] The five-year-old boy setting off on a camping holiday in Figure 13.11 was allowed to cycle unsupervised around the streets where he lived. The absence of traffic and the high population density mean that the streets are overlooked and are full of neighbours walking or cycling. Parental anxiety was refreshingly absent: the skating girls in Figure 13.12 were happy to do another turn for the photo. My host, whose advice I asked before taking these photographs, was surprised to hear how such a request from a strange man in the street would be viewed in Britain. When I read that British children are unhappier and unhealthier than elsewhere,[317] I often think of my time in Vauban.

The recent story of Vauban began in 1992, when the French army left the Vauban barracks to Freiburg City Council. A group of activists formed Forum Vauban to press the Council to create a new carfree neighbourhood there. The Council were initially sceptical but agreed to fund Forum Vauban to consult with the public, who expressed general support for the idea. A list of potential residents – without cars or willing to give up cars to move there – was collected.

Figure 13.11 Vauban children going on holiday with Dad

Figure 13.12 'The kids have taken over the streets'

Andreas Delleske, one of the founders of Forum Vauban, explained:

> "We didn't want to be fanatical about the carfree concept. We wanted to reduce individual car use and offer people the option of carfree living… We wanted the car owners to pay the cost of their own infrastructure…"

Building began in 1998 and finished in 2010. Most of the housing is privately owned: much of it was developed by *Baugruppen*. These are groups of people who come together, often led by an architect or other professional, to commission the building of their own homes, mainly flats, which is usually less expensive than buying one. The car parks are run by a council-owned company and the spaces are not cheap: €22,500 to buy (in 2011),[318] plus a monthly fee to cover maintenance costs. Most of the first residents owned cars to begin with. The extension of the tramline in 2006, the cost and separation of parking, and the culture of carfree living have persuaded many to give up car ownership. The district was originally planned with one space for every two dwellings but many of them lie empty today: car ownership is around 220 per 1,000 people.[319] The spaces can be bought and sold, but many of the owners retain them for visitors or to sell with their properties when they move.

Vauban is not a utopian hippy community: it is like the rest of Freiburg only more so. Even the car owners travel more by other means.[320] The district has faced familiar problems, some of them relating to parking. The district lies on the southern edge of Freiburg, two miles (3 km) or so from the city centre. It borders a village suburb, Merzhausen, where parking is not controlled on the residential streets. A few Vauban residents would park there and claim "that belongs to my grandmother", to circumvent the rules (according to one official I met). The council took legal action against two offenders. Hostility from the media was another familiar problem: when a line of cars was vandalized, for example, the finger was pointed at green extremists instead of the usual, and more likely, suspects.[321]

Stellwerk 60, Cologne

Vauban's *stellplatzfrei* system is an exception among European carfree developments. In other smaller examples, vehicles are physically excluded from the housing areas. Most of them provide some limited parking on the periphery – typically around one space for every five homes, with a few spaces reserved for car club vehicles. There are examples in Edinburgh, Vienna, Amsterdam and several German cities. In some cases (Edinburgh, Amsterdam) the initiative came from the local authority; in others it began with a group of citizens seeking a better quality of urban life. Stellwerk 60 in Cologne began with a *Bürgerinitiative*, a citizen's referendum, calling on the council to allocate a site for a carfree development. Unlike Freiburg, Cologne seemed an unlikely place for such an experiment. Home to the German motor industry, it has a strong car culture and high car ownership, but with the usual differences between the inner districts and the suburbs. After a couple of false starts, some old railway land a couple of miles from the city centre, near to a regional S-Bahn station, was sold to a private developer, with permission to build a carfree development of 400 homes – mainly flats but with 70 terraced houses. Hans-Georg Kleinmann, one of the people behind the *Bürgerinitiative*, bought one of the houses and showed me round as the site was nearing completion.

A path with removable bollards at the entrances runs along one side of the development. A residents' association manages access to the site. The bollards are removed for emergency vehicles, for the minibus for older and disabled people and for big

Figure 13.13 Stellwerk 60, Cologne: shopping by bike

Figure 13.14 Hans-Georg Kleinmann in the communal building

deliveries but not for everyday deliveries (Figure 13.13). Parcels are delivered to a communal building with hand carts and trolleys (Figure 13.14). Each property is allocated between two and five cycle spaces in a mixture of underground and surface lockers: cycle trailers are frequently used for shopping and smaller deliveries. The car club parking spaces are located at the edge of the housing development with a van as well as several cars available. A small multi-storey car park on one corner of the development operates like Vauban's but with fewer spaces – one for every five dwellings.

Figure 13.15 Stellwerk 60: interior

All of this has liberated the area round people's homes for children to play and neighbours to socialise, in the way people used to before traffic and parking took over most neighbourhoods (Figure 13.15 to Figure 13.17). It's not just the kids who benefit: a study in Vienna showed how residents of a carfree development had more friends and knew more people in their neighbourhood than residents of a comparable conventional development nearby.[322]

The marketing of Stellwerk 60 proved a challenge at first. The attitude of estate agents described by Hans-Georg mirrored my own experience, when I was looking to move and give up my car in Britain:

> "The salesmen came along in their Mercedes. People from the initiating group felt the salesmen didn't understand them. It was a bit of a fiasco."

The estate agents initially felt that 'carfree' was a drawback to be played down. They soon learned to better target their marketing, however: both the literature and the website began to sell the idea of a neighbourhood in the heart of the city free from cars, traffic and pollution, a better place to bring up your children. "If we had known when we started what we know now, we could have sold them all without parking," said one of the agents to Hans-Georg, who suggested another unexpected benefit:

> "The title 'Autofrei' keeps certain people away. Those you don't want to see, you don't see here."

Figure 13.16 Neighbours talking in Stellwerk 60

Figure 13.17 A traffic jam in Stellwerk 60!

A preference for or against carfree living was nothing to do with income (30 per cent of Stellwerk 60 was social housing) nor ethnicity (which was mixed): it was a difference of attitude. Think of the most obnoxious anti-social people you have known in your own life: could you imagine them choosing to live without a car to help create a better living environment for everyone?

If this is starting to sound a bit too good to be true, the project did suffer from some of the usual problems – and one rather strange one. When development began, parking was uncontrolled on some of the surrounding streets, which led to the same complaints about overspill as in Vauban. This was being addressed by extending the controlled parking area. The plans to build Stellwerk 60 provoked a curious reaction from some surrounding residents, who organized a demonstration against them. A key objection was that children in the carfree area would be protected from danger, while cars from the small Stellwerk 60 car park would drive down the other people's street, threatening the safety of their own children. Of course, the alternative (a conventional development) would have generated far more traffic, but human reactions to planning proposals are no more rational in Germany than they are in Britain. As a result, relations with surrounding neighbours were strained to begin with, but they improved when both groups united against another planning application, for a railway workshop nearby.

In Britain the term 'car-free housing' – the adjective is usually hyphenated in Britain – usually describes conventional housing with no parking. Car-free conditions are frequently used in inner-London boroughs, which are covered by controlled parking zones. A planning condition attached to car-free new housing precludes the residents from applying for residents' parking permits. These conditions are intended to reduce pressure for parking spaces in dense areas, but they bring no direct benefit to the residents who are subject to this condition.

Although there is no Vauban in Britain, there are a few smaller developments that do offer something more than a 'no-parking' condition. Slateford Green in Edinburgh has 120 dwellings around a central green space and was built by a housing association. There is no allocated parking, but uncontrolled parking is available on some nearby streets and in the car parks of surrounding shops. Bizarrely, tenants are allocated from the housing waiting list with no mechanism for choosing people without cars.[323] People at the top of housing waiting lists don't tend to own many cars, but those who do are able to park them nearby. These minor problems aside, the development seems to have worked reasonably well.[324]

Another type of car-free development is so familiar that it passes almost unnoticed. Pedestrianization has traditionally been viewed in Britain as a means of increasing retail spending – a successful one according to most studies.[325] Some planning authorities assume that people can't live where cars can't drive, failing to notice that many already do. Some developments like the Brunswick Centre in Bloomsbury (Figure 13.19) have been around for many years. Others, on Bristol's harbourside and Exeter's Princesshay (Figure 13.18) have been built more recently. Developers often assume that parking and vehicular access are essential to sell properties but in pedestrianized central areas this is rarely true. Exeter's Princesshay redevelopment included 120 flats with just 23 parking spaces: people queued in the street overnight to buy flats in the first phase, which had no car parking. The car-free flats on Bristol's harbourside also sold with no difficulty. I looked at one of these flats when planning our move to Bristol, the year after my visit to Cologne: unfortunately it did not occur to whoever planned them that residents without cars might need secure space to store bicycles!

So if more imaginative developments like Vauban or Stellwerk 60 were built in Britain, who, if anyone, might be willing to move there? (That, by the way, was the main question my PhD set out to answer.) The greatest potential, at least in the short term, is likely to come from 'car-free choosers' – people who live without a car by choice. There are more of these people than you might imagine. Among adults without a car, only a minority say they can't afford one. A wide range of other reasons can be found in different surveys, but underlying many of them is the simple observation that in some places people feel they don't need a car.[326]

Figure 13.18 Flats in Exeter's Princesshay

Figure 13.19 Brunswick Centre, Bloomsbury, London

Three-quarters of the car-free choosers in my sample lived in cities of 100,000 people or more, concentrated particularly in the inner areas. They tended to like the vibrancy of urban life but not the traffic and pollution. They were slightly younger than average overall, although nearly half of them were over 40. After many years of decline, the proportion of households without cars has stabilized at a quarter since 2005.[327] But, as we saw in Chapter 6, that stability masks some interesting dynamics. In any one year many people acquire a car and a similar number give them up, for a wide range of reasons: retirement, separation, moving to a more urban area[328] or because an old car finally fails and they decide not to replace it.[329] Car ownership has been falling in our cities and rising elsewhere.[330] It has been rising among older people, as the first generations who grew up with car ownership reach retirement age, and falling among young people, not only in Britain but also across the developed world.[331] The average age of passing the driving test and acquiring the first car has been rising since the mid-1990s.[332] The reasons for, and implications of, these trends have generated much academic debate.[333] The trends are too recent to allow us to draw any firm conclusions but we do know that people who pass their test later in life drive less throughout their lives than those who start young.[334]

Whenever I talk about carfree development or carfree living I am often confronted with incredulous questions like "But how would you do your weekly shopping?" as though the quarter of households without a car were about to starve to death because they couldn't make it to shops! Carfree living is not a utopian ideal; it has always been with us. It is growing in our cities and among younger generations and it offers an opportunity for the future. If the past two chapters have given the impression that progress is only happening in continental Europe, the next two will describe positive changes in some British cities, starting with London, which, in terms of transport and many other respects, is in a category of its own.

CHAPTER 14
London: the politics of bucking the trend

"No one's interested in transport itself, apart from engineers and enthusiasts. Transport is the means by which a major city works. People look at this as a transport question without recognizing the political conditions you need…"

Peter Hendy

Peter Hendy, commissioner for Transport for London under mayors Ken Livingstone and Boris Johnson, explains why London succeeded where most British cities have failed. Figure 5.4 (page 44) showed how bus use in London has bucked the UK national trend, doubling since the mid-1990s. With walking as a principal mode of travel remaining stable, the main story told by Figure 14.1 is a progressive switch from driving to public transport. Cycling in London has also nearly doubled in recent years, though starting from a very low base. Cycling is important in parts of inner London but across the city as a whole just 2 per cent of journeys are cycled. Total traffic *volumes* (as opposed to percentage shares) reached a peak in 1999 after which they began to decline, slowly at first, but more rapidly from the start of the recession in 2008 despite a strongly rising population.

Transport has always been highly politicized in London. It is "an embarrassment to central government", says Hendy, "because it costs a lot of money" and because parliament and government ministries are surrounded by it. Conflict between central government and London's leaders came to a head over transport issues in the early 1980s.

In 1981, a leftwing Labour group under Ken Livingstone took control of the Greater London Council (GLC). Dave Wetzel (a firebrand councillor judged "too leftwing" to look after Hounslow's burial committee "in case I went round the gravestones with Karl Marx, raising the dead") was put in charge of transport. The group was young, lacked experience and many people doubted their ability to implement the big changes they were promising. Years later, one of the officers (the council's employees) told Wetzel, "I

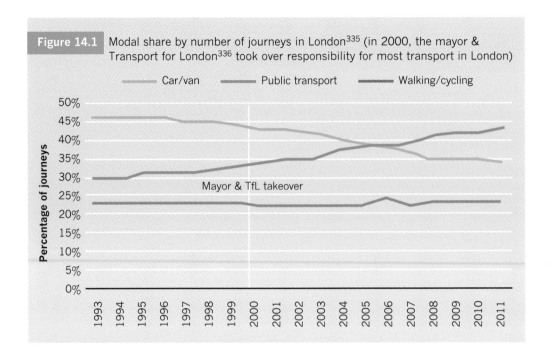

Figure 14.1 Modal share by number of journeys in London[335] (in 2000, the mayor & Transport for London[336] took over responsibility for most transport in London)

said to my lads: give 'em six months and everything'll be back to normal, and here we are five years later… Things never did go quiet or go back to normal."

Conflict with Margaret Thatcher's government was one of many problems they faced. Though planned overspill to new towns had come to an end, London was still haemorrhaging people. People who don't remember those times may find it difficult to imagine how attitudes to urban living have changed, particularly among young adults.[337] I moved to work in outer London at the age of 22, determined like many of my peers to buy a cottage in the country as soon as I could afford one. (Being able to afford one might be difficult to imagine nowadays!)

Opposition from the public and the Treasury had killed the big plans for urban motorways in London but the first of Buchanan's options, to knock down and spread out, was still influential. A big programme of road building was still part of GLC policy when Wetzel took over. "We consulted on all the Tory [Conservatives'] road schemes and scrapped the vast majority of them", apart from three big schemes in Hayes, Tottenham and along the A2, "because Ken [Livingstone] had been to those areas promising to build bypasses." He adds, "Without them, we might not have won the election."

At the same time the GLC's political leaders stopped the officers from working up new schemes for any future administration:

CHAPTER 14 London: the politics of bucking the trend

> "We said, 'These are the schemes you're gonna build. **These are their reference numbers and if you're working on any scheme which hasn't got that reference number you're out the door.**'"

Officers pursuing a different agenda would remain a problem for Wetzel at the GLC and also when he returned years later to work for the mayor. The previous administration had tried to address the decline in public transport with plans for a zonal travel card, but the plans were opposed by London Transport:[336]

> "The officers kowtowed too easily to external forces... **Now we know London Transport was totally wrong. Luckily the leadership had changed when we took over.** A zonal system with integrated ticketing was part of our manifesto. London Transport was ready for that, but not in one go."

A fares system based on zones was introduced in 1981; the 'Travelcard', a season ticket covering buses and the underground, followed in 1983 and a one-day Travelcard was introduced in 1984. In between came one of the fiercest conflicts between central and local government over a transport issue. Wetzel recalls:

> "In 1981, David Bayliss, chief transport planner [an employee of the GLC], presented me with [a report which] said 'there is a secular decline in the use of public transport'. **I said: 'I don't care if it's secular or religious. We are going to get more bums on seats.'** We immediately took off an overtime ban (imposed by the management to save money), started operating better services and started consulting on lower fares."

Wetzel argued for free public transport, but settled for a cut of a third in average fares. The subsidy required to cover the cost of the cut was mainly financed by an increase in domestic and business rates (which were still set locally in 1981).[338] 'Fares Fair', as the policy was named, proved popular with most of the public, but the Conservative-controlled borough of Bromley objected to subsidizing the underground, which did not run through their borough.

Legal arguments followed (and went all the way to the House of Lords) about the interpretation of 'fiduciary duty' ie the duty to avoid wasting public money. Was the subsidy consistent with this duty in the Greater London Act (which specified the powers and duties of the GLC)? The "vandals in ermine", as Wetzel dubbed the law lords, decided it was not.

The leftwing Labour councillors, including Livingstone and Wetzel, voted to ignore the court ruling, but in March 1982 a majority of the councillors voted to double fares.

Figure 14.2   Dave Wetzel in court in 1982 (drawn by the court policeman)

Wetzel and a few of his colleagues organized a campaign, 'Can't Pay, Won't Pay', which landed him in court for non-payment of fares. "I borrowed a ventriloquist's gorilla from the National Theatre," he says with a smile, "'cos the magistrate at the committal hearing said I was behaving like an animal." (Figure 14.2)

The lower fares had lasted for six months, just long enough to reverse the decline in public transport and start reducing traffic flows. That all changed when the fares went up: public transport of all forms hit a long-term low in 1982. Then something quite unexpected happened.

Several times in my interviews with them, Wetzel or Hendy would say "you ought to ask Ken Livingstone that". A few weeks later I meet Livingstone in the modest terraced house in north-west London where he still lives. ("Pensioner, house husband" is how he describes his occupation now.) His wife's Toyota Prius sits outside, but Livingstone himself never learned to drive. "I was born in 1945… it was too expensive back then," he says, adding that he prefers to read while travelling and thinks of all the time he would have wasted sitting in traffic jams if he had been driving a car. With ambition behind him, he reflects on the past with the cynical wisdom of an idealist who has tested the limits of what he can change. I ask if I can record and he says: "Why not? This place has probably been bugged for years."

From 1983 onwards, all public transport, but particularly the buses and underground, began to recover (Figure 14.3). Why was that? The Travelcard introduced in 1983 brought some of the simplicity observed in Freiburg (though in London the daily version, introduced in 1984, could only be used off-peak). It also reduced the fare level. How were the GLC able to get away with this so soon after the law lords' ruling?

> "I received informal word from the law lords [via the GLC's lawyers] that if we cut fares again they wouldn't overturn it..."

The original decision, he and Wetzel believe, was political:

> "They wanted to constrain this dangerous radical – the enemy within. Normally when the law lords make a reactionary decision it only affects a handful of people, but this time... I think they realized they'd made a catastrophic mistake."

By 1983, the Thatcher government had had enough of 'the enemy within': it announced plans to abolish the GLC. Eager to wrest control of transport away from Livingstone, as a first step to abolition, it created London Regional Transport (LRT) with a board which it appointed in 1984. Transport in London was, as Wetzel points out with some irony, "nationalized".

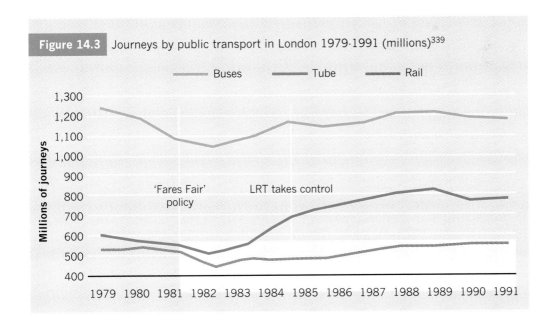

Figure 14.3 Journeys by public transport in London 1979-1991 (millions)[339]

As shown in Figure 14.3, from the mid-1980s, all forms of public transport in London began to stagnate, with buses and the underground declining after 1988. Would it have made any difference if the GLC had never been abolished? "Not much," according to Livingstone.

> "One of the first laws of the Thatcher government was to take control of capital spending. **So even if we had a Labour-controlled GLC, we couldn't have borrowed money to carry on expanding the transport network...** Mind you, you would have at least had a voice making the case for it, and we would have been ready to start stuff in '97 when Blair got in – but not dramatically different."

London Regional Transport sold off its bus operations, to management buyouts that were soon bought up by the big bus companies. But full deregulation was postponed. (I remember an eager young radio journalist at the time asking an official when Londoners could also enjoy the benefits of deregulation. With some of my family enjoying these 'benefits' in other cities I had a quiet chuckle. I wonder what the journalist would think of his question today.)

One of the sticking points with deregulation was the Travelcard, which had become very popular. John Major's Conservative government set up a committee to work towards deregulation. The private operators told the committee that the Travelcard would have to go, unless its price was substantially increased, or the taxpayer subsidized the difference. The committee's report was then leaked to Christian Wolmar, transport correspondent of *The Independent*, who asked the 'Save the Travelcard' campaigners, for their views on it. Their spokesman summed up the attitude of many people:

> "This report is a catalogue of disasters for London. **If this is to be the result of privatisation and deregulation, then these changes will have failed completely to meet the needs of Londoners.**"[340]

The Travelcard issue was about more than just finance. Buses used to have a conductor who collected the fares. During the 1960s and 1970s, most buses in London and elsewhere gradually switched over to 'one-man operation', where drivers had to collect the fares. The queuing to pay the driver increases journey times and surges of passengers boarding at particular stops can reduce the reliability of timetables (still a problem on buses outside London today). The prepaid Travelcard significantly reduced these delays. As the Major government struggled through scandals and national economic problems, the plans were quietly dropped.

The mayor of London and Transport for London

The question of London's governance would not go away. In 1997, the Labour party won the national election with a manifesto which included a commitment to create an elected mayor and city-wide authority for London. This was about more than party politics. According to Hendy:

> "Business got fed up with the incoherence of a major city run without a strategy. It was London First [a business lobby group] and the CBI who said: 'We can't live with this anymore. We've got to have a strategic authority.'"

The national leadership of the Labour party did not want a radical like Livingstone returned to power, so he stood as an independent and was elected as the first mayor in 2000. Though the business community might have preferred someone else, even a return of their old enemy was a price worth paying for London-wide coherence. The Corporation of London, "a front for the Tory party" in the 1980s, had changed its views and lost some of its influence to groups like London First, who worked more with Livingstone. Ken was a "good socialist politician," says Hendy, but he and the business community were "almost wholly aligned" on transport issues "as was Boris".

The new mayoral structure was smaller than the old GLC and less concerned with directly running services, but it did take over London Regional Transport – renamed Transport for London (TfL) – with responsibility for buses and major highways in London. The underground remained under national government control until it too was transferred to TfL in 2003. I asked Livingstone whether he had any political battles over transport issues and he said:

> "There weren't any battles because the decision was solely mine. That's the weakness of the mayoral system – far too much concentration of power... I achieved just as much as the leader of a slightly dysfunctional Labour group on the GLC..."

Two of the challenges for the new mayor were to produce a London Plan and a London Transport Strategy. The London Plan set the context for the Transport Strategy and everything else that the mayor and the other authorities in London would have to deal with. The key issue was how to accommodate the growing population. As shown in Figure 14.4, the post-Second world war decline in London's population went into reverse from 1988, with the increase accelerating in more recent years.

Overspill or encroaching on the greenbelt were never serious options: London would have to live within its existing boundaries. For Livingstone this was no real issue.

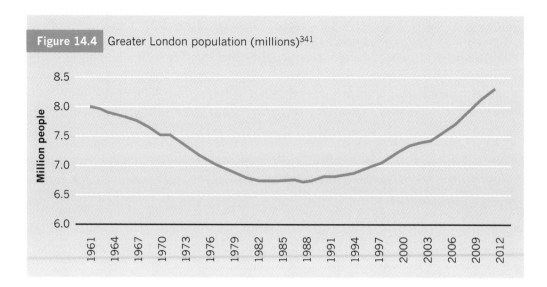

Figure 14.4 Greater London population (millions)[341]

Paris and New York are both much denser than London, he claims. Comparisons of this type depend on boundaries. The greenbelt has moulded London into a single, almost circular conurbation. The city of Paris, which is denser than London, is tiny by comparison. It is surrounded by suburbs that are irregular in shape, as they are around New York. The City Mayors Foundation, which attempted to measure boundaries on a like-for-like basis, showed London in 2007 as the fourth densest large city in Europe, much denser than Paris and two-and-a-half times denser than New York.[342] How do you plan for the transport of a growing population in such a city?

On this issue, all the people I interviewed agreed: accommodating the increasing population by increasing road capacity would be impossible. As Livingstone put it:

> "In great world cities like London and New York with eight million people, people have to use public transport…**It was quite obvious by the Second World War in New York, whatever road capacity you put in, it would fill up with cars.** It was particularly stupid that post-war Labour and Tory governments came up with the three ringways for London…the cost of building motorways through London was absolutely horrendous, even back then. **Nowadays with [much higher] land values, it just couldn't be done.**"

Livingstone published his Mayor's Transport Strategy in 2001. This was restrained in its language but clear in its ambitions: [343]

- Reducing traffic congestion through a congestion charge.
- Overcoming the backlog of investment on the underground, increasing capacity by 50 per cent over 15 years.
- Making radical improvements to bus services.
- Better integration of National Rail with London's other transport systems.
- Increasing the overall capacity of London's transport system (mainly rail, but also an east London road bridge).
- Improving journey time reliability for car users, while reducing car dependency by increasing travel choice.

Although local transport schemes, walking and cycling get a mention, the emphasis was clearly on public transport investment. Apart from the congestion charge in the city centre (see page 174), there was no declared aim to restrain travel by car and there would also be no repetition of Fares Fair. In 1981, the underground had empty seats waiting for bums; by 2000, the problem was overcrowding. Increasing capacity would take time and money. In the meantime, buses were the first priority.

Improvements to London's bus services

When Livingstone took over, London's buses were almost breaking even, apart from concessionary fares. However, the poor quality of service was beginning to affect a labour market recovering from the recession of the early 1990s, particularly in inner London where lower-paid staff were becoming difficult to recruit and retain. This was one of several issues where the new team was determined to change the culture of the organization it inherited. "The bureaucracy had solidified," says Livingstone. The day before TfL took over, and again when it gained responsibility for the underground in 2003, senior people took early retirement. Wetzel remembers a letter he received from the managing director of London Buses urging him to:

> "look at the tendering system because it was an oligopoly in decline and we should bring in some fresh new tenderers who would compete on price. I went and saw Ken and said: 'This bloke don't get it. We want to improve the quality. We want to pay more, not less, for our bus services.' As a result of that letter, that man took early retirement. Some people at London Buses never forgave me for that."

The managing director's eventual replacement was Peter Hendy, who had led one of London Buses' management buyouts before joining First Group. With more money available, Hendy increased routes and frequencies among many other improvements.

> "There's a common misconception that [the cost of subsidizing a good service is] about evenings and weekends. **It's not**. What you're paying for is a peak bus service that carries the last set of people at a bus stop to work."

Lianne De Mello, a council officer in Brighton and Hove, whom I was interviewing for the next chapter, was brought up in Harrow in outer London. She recalled:

> "I remember pre-Ken, about the time I was becoming a teenager, looking to get out and about without my mum and dad, I remember it being difficult to get around. **Then suddenly there were buses everywhere...** I think there's a psychological difference between having to look up the time of a bus, or just turning up at the stop...nowadays, **unless it's a Sunday or a bank holiday you never have to wait more than 10 minutes for a bus.**"

Passenger bus miles increased by three-quarters during the eight years Livingstone was in charge.[344] The introduction of the 'Oyster card', the 'contactless' smartcard which replaced cash for most passengers, in 2003 was another important element. Today, the Oyster card can be used for season tickets, or it can be used on a pay-as-you-go basis, in which case, you will never be charged more than a daily maximum rate. Thus it gives customers either single fares or the daily rate (lower than the printed Travelcard), whichever is cheapest. It is valid on almost all forms of public transport in London, including river boats and national rail services. It wasn't like this at the beginning.

The original Oyster card contract had been let by London Regional Transport before the mayor took over. The Oyster card was to cover buses and the underground, but not the Docklands Light Rail (DLR), the Croydon Tramlink or national rail services in London. The contractors also wanted to restrict it to monthly and annual passes. The Board of TfL was frustrated by this attitude. Livingstone's advisers called on their connections in Moscow, to line up an alternative provider if the contractors refused to cooperate.[345] The pay-as-you-go and Travelcard elements were incorporated at its launch in 2003. Then, Wetzel explains:

> "When the private company failed on some of its targets, instead of giving them a cash penalty, we said: 'You have to include the Croydon tram and the DLR.' **When London Transport let the contract there were ongoing, ongoing, ongoing talks about a national smartcard.** They felt it was going nowhere and they would still be talking about it in 25 years' time. **So they pulled the rug under the discussions and went their own road...**"

In 2010, the Oyster card was extended to national rail and river services; then in 2014, with operators in the rest of the UK still struggling to implement any smart ticketing, TfL introduced another innovation. The ordinary contactless credit and debit cards issued by most banks can now be used in the same way, benefiting from the same fares as the Oyster card. Cash handling on London's buses has now been withdrawn altogether.

The London underground

After many years of under-investment, the underground badly needed renovation and more capacity, which was going to be expensive. The government had decided to transfer ownership of the track, stations and auxiliary services to a public-private partnership (PPP) before the mayor took over. TfL would only operate the trains and employ the station staff. The PPP idea emerged as a compromise between Chancellor Gordon Brown, who wanted full privatization, and Deputy Prime Minister Prescott, who disagreed.[346] Livingstone recruited Bob Kiley as transport commissioner. Kiley had transformed the public transport systems of Boston and New York, and brought his own team of seven or eight Americans with him. With the 'dead bureaucracy' removed, Livingstone and his team believed TfL could do a better job in-house. Wetzel recognized that the past management of the underground had not been good, though the Treasury, with its 'on–off' investment funding, had been part of the problem:

> "From what I heard, the old underground management team supported PPP not because it was the cheapest, most efficient way of renewing the underground, but they knew, once they had contracts signed for 25 or 30 years, the Treasury was bound to that contract – they couldn't switch it off…"

PricewaterhouseCoopers produced a report showing that PPP would be a cheaper option but there were good reasons for treating this idea with scepticism. The consortia that won the two contracts would have to borrow at higher interest rates than TfL could obtain. Instead of letting the contracts on a 'job-by-job' basis as TfL would do, the consortia would sub-contract to their member companies, often at higher costs and with no penalties for any underperformance. Other problems, according to Wetzel, were due to the inherent limitations of contracts:

> "We knew that once a contract had been signed it would be inefficient because everything that was not in the contract would be charged over the odds. **How do you define in a contract what is acceptable or unacceptable litter on a station?**"

Attempts to negotiate with central government foundered on the opposition of Chancellor Gordon Brown so eventually TfL took over the underground with the PPP in place. TfL found itself paying 'over the odds' for changes not specified in the contract, like re-siting signals for safety reasons. Four years later its worst fears were realized as the Metronet consortium ran into financial difficulties. Livingstone told everyone to avoid any hint of 'I told you so' while he negotiated with the Treasury to take the contract back in-house. The National Audit Office concluded:

> "The Metronet PPP contracts to upgrade the Tube left the DfT without effective means of protecting the taxpayer. Metronet's failure led to a direct loss to the taxpayer of between £170 million and £410 million."[347]

A few years later, under Boris Johnson, the remaining PPP consortium went the same way. After serious problems with contracts let by central government for extensions to the Jubilee underground line and the DLR, the Treasury finally allowed the mayor and TfL the freedom which most European mayors would take for granted – to borrow money and manage the contract for building Crossrail, the next big transport infrastructure project. Crossrail would provide London with an east–west suburban rail line in a tunnel, similar to the RER line A, which opened in 1977 in Paris. It was one of many transport projects at different stages of planning or implementation when Johnson took over from Livingstone in 2008.

The London congestion charge

The most controversial transport policy initiated by Livingstone was the congestion charge, a daily charge for vehicles entering central London. As he points out in his autobiography, the idea came from the political right. For neoliberal economists the logic is straightforward: a scarce resource should be allocated to those who pay for it. For Livingstone, who made it a manifesto commitment, it was the only way to reverse growing congestion and pollution in central London.

The run-up to its introduction was accompanied by a barrage of opposition from the media, celebrities and ministers in the government which had created the power to introduce congestion charging, in the Transport Act three years earlier. Much of this opposition was personally directed at Livingstone. "Ken's off his rocker," said Michael Winner to *The News of the World*. *The Guardian* said ministers planned to "accuse Livingstone of incompetence" if it went wrong.[348] Even Livingstone's advisers at City Hall had second thoughts, urging him to delay until after the next mayoral election. Wetzel recalls:

> "Ken came over to TfL and discussed this with Kiley, who had reviewed the plans and was sure we were on the right track. He said to Ken: 'I promise you, if for any reason you're not satisfied, I can switch it off six weeks after you give me the instruction…' and it was like a safety blanket for Ken."

Livingstone says his personal commitment never wavered, but that of many around him clearly did.

The technology used to charge drivers for entering the charging zone (based on number plate recognition) was not the most up to date, nor the cheapest to run – it was chosen to minimize the risk of a spectacular failure, but administration costs absorbed much of the revenue. On the day, the media predictions of mayhem were proved wrong and other cities such as Stockholm and Milan later followed London's example by introducing their own congestion charging.

Cycling in London

One indirect impact of the congestion charge was a boost to cycling in central London. Cycle counts on TfL's roads doubled in the seven years following its introduction. This increase was from a very low base and concentrated in central areas. One borough, Hackney, stands out: the 2011 census shows 15 per cent of commuters normally cycle to work, twice as many as inner London as a whole, and the gap has widened since 1991. Why was this? Trevor Parsons of the London Cycling Campaign ascribes it partly to demographic change, though similar changes were happening in some other boroughs. Councillor Vincent Stopps, who took charge of transport there in the early 2000s, said that "restraining the private car is key". Many of the measures they took – widening the area where parking is controlled, the car-free developments – were common to most inner-London boroughs, but the difference, he argues is:

> "We know what we're doing here. In some of the neighbouring boroughs the engineers take a design out of the design manual and mess up the details… We don't do 'cycling schemes': we do public realm schemes that benefit cyclists, walkers and the bus."

Hundreds of mainly small-scale changes were central to their success. Parsons, whose local group of the London Cycling Campaign is influential in Hackney, took me on a ride around many of them. From the 1990s onwards, the Campaign took a strong stance against low-quality infrastructure. Instead of paint on roads, stop–start cycle lanes and shared pavements, they have pressed for, and obtained, traffic calming, junction redesigns, and filtered permeability. Many residential and side roads were

Figure 14.5 Trevor Parsons at filtering point in Hackney.

Figure 14.6 Cycle contraflow in Hackney.

closed to through traffic some time ago; they needed only minor changes to create cycle-friendly filtering points like the one in Figure 14.5. Elsewhere, new filtering points were installed or contraflows introduced on one-way streets (Figure 14.6). The cumulative effect of all these changes is to criss-cross the borough with quiet roads (and some paths) which favour cycling and walking.

The demographics and behaviour of London's cyclists are very different from the cities described in Chapters 7 and 12. Nearly three-quarters of journeys taken by bicycle are made by men.[349] On the Barclays Cycle Superhighways, which are mainly painted lanes on roads, including many main roads, just over three-quarters of cyclists are male. A study of the Barclays Bike Hire scheme showed that women tend to avoid main roads and to use the bikes more for recreational riding in areas like Hyde Park.[350] In *The mayor's vision for cycling* (mentioned in Chapter 7, page 77) Boris Johnson declares an aim to get more women cycling, through the construction of safer cycle routes.

London transport policy under Boris Johnson

Livingstone is scathing about his Conservative successor Boris Johnson, on cycling policy among other issues:

> "We were working on the next stage of this and our cycle routes were going to get as much physical separation as possible. **Boris just dumped all that and painted blue lines on the road, because Boris never wants to offend the car lobby...** That would have been the biggest transport part of my third term if I'd had a third term... **We were**

going to close the north side of the Embankment between Westminster and Waterloo Bridge every summer, replicating what the mayor of Paris did. We were going to do in Parliament Square what we did in Trafalgar Square, close one side. There would have been many more real shifts towards pedestrians and cycling."

Hendy agrees that, "Ken was more keen on segregation than Boris", but in most respects he believes there has been more continuity than radical change in transport policy:

"Look at the names on our board prior to and after Boris's election. Some of the names are the same. He was bound to get rid of Ken's political lieutenants [including Wetzel] but interestingly, he didn't sack the two trade unionists on the board… Boris, to his credit, has not changed the bus network… [Conservative Board member] Stephen Norris came back with a vengeance in 2008, saying it was inefficient, cost too much money; we ought to cut things, change things. We got a report written by KPMG who said it was well planned; it did what was asked of it and Boris buried it."[351]

Livingstone criticizes Johnson for abandoning many of his investment projects. Although London fared better than most of the country it was not immune to the financial downturn. Hendy's view is:

"If Ken had won in 2012, he wouldn't have been able to put the fares up so much and I suspect we wouldn't have done so much investment."

Many of the largest investment projects under way will increase capacity on the underground. Even as bus use was soaring, the underground increased its passenger miles by 14 per cent during Livingstone's tenure.[352] By the late 2000s, many stations and some sections of the network had reached their capacity, with entry to stations sometimes closing at peak times.

Towards the end of Livingstone's time, TfL took over several poorly used rail lines in north London, creating the London Overground. The policy of upgrading and expansion has continued under Johnson, with substantial increases in patronage – the original north London lines more than doubled passenger numbers in the first four years of TfL operation.[353] Hendy would like TfL to take over more of the suburban rail lines, particularly in south-east London, which is not served by the tube. Resistance from Kent county council has prevented this so far.

Explaining the changes in London's travel patterns

Ben Plowden, TfL's Director of Strategy and Planning, is one of the new generation of senior managers at TfL. He has been promoted since I last met him and talks with conviction about the "world class" organization he represents. How does he explain the modal shift achieved in London since TfL took over? His answer is carefully phrased:

> "There is a helpful mix of politics, provision and attitude…the first mayor of London had a very explicit agenda to encourage and enable people to use their car less…the second mayor has sought to differentiate himself in being more balanced between modes but has not engaged in a major programme of road building. There has been a clear political commitment to supporting sustainable travel, in Boris's case particularly (though not exclusively) the bike. Secondly, the physical and transport choice environment makes it easier for people who could afford a car not to do so…"

He speculates about demographic changes in the capital – a young population, people moving from countries with less of a car culture – and adds that institutional arrangements are fundamental. TfL has control or influence over all forms of transport in London, and the mayor also controls strategic planning. So what difference does that make in practice?

> "If the tube goes down [we can] tell people in real time about their alternative options. How is Crossrail going to connect along its route? What cycle parking should we put at Euston if HS2 goes there? How do you connect buses at local town centres to the increasingly used Overground network? Those are questions which most transport authorities can't even ask because they don't own the answer [but TfL can, because of the breadth of its powers]."

Like Hendy, Plowden stresses the positive work of TfL in changing London's travel patterns. But how much of the change is due to deliberate policy, and how much of it would have happened anyway? A key issue is capacity, in terms of roads to carry vehicles and in terms of parking space. As with many other transport issues, indirect or unintended effects may be at least as important as intentions. London is Britain's second densest city, after Portsmouth. Its population, and density, rose by a quarter between 1988 and 2013 (see Figure 14.4). The rise was gradual at first but began to accelerate from the late 1990s. At the same time, the capacity of London's roads was declining. TfL estimates the capacity of its road network by looking at the volume and speed of traffic. For any given volume, falling speeds indicate a fall in capacity. It is a rough and ready method but gives a clear picture of the direction of change. In central London, road capacity fell by a third between 1992 and 2009. The falls for inner and outer London were 15 per cent and 5 per cent respectively.[354] Plowden explains:

> "Over the past 20 years, many local incremental changes have been made to the network, almost all improving the value of places and giving priority to modes other than the private car, so pedestrian crossings outside schools, bus lanes, cycle lanes, road safety improvements, public spaces... Local communities have got these small-scale, sometimes large-scale improvements to the highway environment."

All of which has reduced the capacity of the network to carry traffic. Parking has been another area of major change. London was the first British city to introduce parking controls from the late 1950s. The process accelerated during the late 1990s, with the spread of controlled parking zones, which now cover most of inner London. Estimates of total parking capacity are difficult to obtain, but one study estimated that uncontrolled on-street parking fell by 5 per cent between 2001 and 2010 across London, with the decline in inner-London boroughs ranging up to 44 per cent (in Islington).[355] As the controlled parking zones spread, several inner-London boroughs have introduced car-free housing policies. The impact of these policies was limited at first, but has been growing over time.

The trends in car ownership (Figure 14.7) have been dramatic. At the time of the 2001 census, car ownership in London had, surprisingly, risen to around the national average. Over the next decade it fell by 27 per cent – more in inner London, where only 39 per cent of households had access to a car by 2011.

Apart from population density, demographic changes offer no clear explanation for these trends. The proportion of people in employment and people with children, both of whom tend to own more cars, increased over those 20 years. On the other hand, there were more young people, who own fewer cars, and fewer homeowners, who own more. There are some interesting differences in the travel patterns of ethnic groups: the 'white British' group own more cars and drive the most, partly because more of them live in the suburbs.[356] Whether the changing ethnic composition of London has contributed to falling traffic levels is an interesting question which remains to be studied.

A few years ago, I did some research into households without cars in Bloomsbury and King's Cross – two areas with particularly low car ownership. When asked why

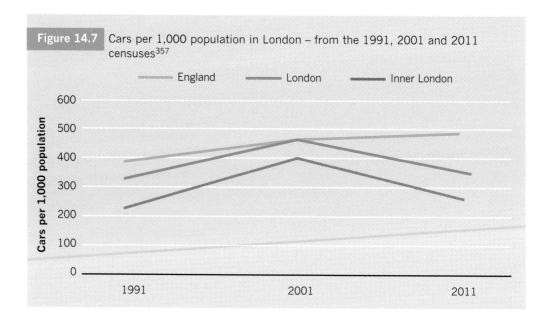

Figure 14.7 Cars per 1,000 population in London – from the 1991, 2001 and 2011 censuses[357]

they did not own a car, cost was not the main reason: most could afford one if necessary. The most common reason, cited by just under half, was "I have no need for a car". Lack of parking was rarely the main reason, but was a secondary reason for just over a third of them.[358]

How much difference the congestion charge made is difficult to assess. The last monitoring report showed traffic volumes 16 per cent lower than 2002. The composition of this traffic had changed: cars had fallen by 36 per cent, vans by 13 per cent and lorries by 5 per cent, partly replaced by more buses, taxis, bicycles and motorcycles.[359] Although traffic volumes have fallen, initial reductions in congestion were not maintained. One study estimated that road capacity fell by 20 per cent between 2004 and 2008.[360] The pedestrian improvements mentioned by Plowden are a big part of the reason for this.

Another study argued that the congestion had little impact on modal shift, most of which came from improvements to bus services. It also pointed out that Birmingham achieved similar traffic reductions in its city centre without a congestion charge (though the increase in Birmingham's population has not been so rapid).[361] Traffic in London's congestion charge zone is a very small proportion of the total, so its direct impact on London as a whole was never going to be very great.

The cost of transport in London

The 'tried and tested' technology chosen for the London congestion charge was expensive to operate. In the early years the net revenue made very little contribution to the expensive subsidized buses, but that has gradually changed, as costs have been controlled, and bus fares raised. By 2012-13, the congestion charge was contributing over half the cost of subsidizing London's buses.[362] As in Freiburg, increasing subsidies from 2002 onwards helped to stimulate bus use, which continued to rise as the subsidies were reduced. Between 2007-8 and 2012-13, the cost of subsidizing London's buses fell by nearly a half.[363]

Notwithstanding those economies, in 2012, London was still spending over twice as much per person as the other English regions on both capital transport projects and subsidies to running costs. In response to critics in the north, a parliamentary scrutiny committee pointed out that most of this money is spent by local authorities or other public bodies – not central government.[364] This is not a particularly convincing rebuttal. The difference in local authority spending reflects a more generous central government settlement for TfL than for local authorities or transport authorities in other cities. London's unified political structure is clearly a major advantage. The moves towards combined authorities (controlling highways and some other powers as well as public transport) in some of the northern conurbations is a step in the right direction, although they will still lack the regional planning powers available to the mayor of London.

Regulation of buses and control over suburban rail networks are other key advantages that London has. Livingstone believes deregulation would be impossible in London: apart from the effects of "a crap service" on the functioning of the city "there isn't the room on our roads for competing buses". Hendy put it rather ironically as follows:

> "Any city in Britain can have the system in London, if it'll pay for it… **I've run bus services in provincial Britain; I can run them very successfully.** You put the peak fares up, price people off so you haven't got a big vehicle requirement. It's economical to run, works a treat. **What it doesn't do is take everybody to work at cheaper fares.** Now if that matters to you as a city, you can do something about it."

All of this presupposes the ability to raise money and overcome the opposition of vested interests like those described in Chapter 5. Central government has devolved more power to raise money for capital projects, but has tightened its grip on current spending. Hendy cites Manchester as a city which has used the system to its advantage to build a large and growing tram network. This has undoubtedly benefited

Manchester, but as we saw in Chapter 5, has made little difference to overall patterns of travel in Manchester.

Both Hendy and Livingstone emphasize the importance of business attitudes. Unlike many other cities, business leaders in London understand that accommodating everyone who wants to drive is impossible, and supporting alternatives is essential.

Observations and conclusions

As in Lyon, the big investments in public transport in London have been relatively expensive. Although the early investment in buses is now allowing subsidies to be reduced, in the future a more determined strategy to promote cycling could prove more cost-effective. It is sometimes argued that cycling can only make a big contribution in smaller cities. The cities with the highest rates of cycling are all much smaller than London, but this may be at least partly coincidental: the European countries with the highest proportion of cycling don't have any cities as big as London. Cycling in Tokyo accounts for 14 per cent of journeys compared to 2 per cent in London.[365] Hackney, the London borough with the highest rate of cycling, has made filtered permeability a guiding principle, providing separation by closing minor roads to through traffic.[366] The 2013 policy document, *The mayor's vision for cycling* promises a break with the past. Plowden says the funding for implementation is secure and they are serious about making it happen. Will the difficult decisions that are necessary to deliver this vision be taken or fudged? Time will tell.

Time will also tell whether a new roads strategy will deliver improvements to the public realm or consolidate the grip of motor traffic on the city. The aim of the Roads Task Force was to reconcile essential movements with pressure to improve the pedestrian and cycling environment. The approach of its report seems to hark towards Buchanan, with some roads designed to move vehicles as easily as possible, and others, like Covent Garden or the City of London, to create "a world class public realm".[367]

Despite the incremental pedestrian improvements, walking around London still doesn't feel as pleasant as in many continental cities (and some areas still suffer from dangerously bad air quality[368]). Hendy says that's inevitable in a larger city, and there may be some truth in that. But strolling along the newly pedestrianized bank of the Seine in Paris a few months later (Figure 14.8) reminded me of Livingstone's unfulfilled plans for Westminster Embankment (Figure 14.9).

In the years since Christian Wolmar wrote about the threat to the Travelcard, he has become one of the UK's leading transport writers. His books lift the lid on the failings of privatized rail and deregulated buses. In 2013, he joined the competition

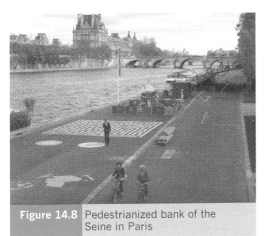

Figure 14.8 Pedestrianized bank of the Seine in Paris

Figure 14.9 Westminster Embankment

to be selected as the Labour candidate for the 2016 mayoral election. If elected, he explained, he would set out a long-term vision for transport and planning. For inner London, the strategy would reduce vehicle movements, "reclaim road space in order to create a better environment" and encourage walking and cycling. In outer London, he would intensify around suburban centres and flatten public transport fares "so those who are forced to live further out aren't penalized". It sounded like a programme to continue where Livingstone left off.

Most politicians fall foul of the media at some point, but those who buck the political trend can expect no mercy. Livingstone's autobiography recounts a litany of persecution by sections of the media. He has been accused of everything from insanity to supporting terrorism. Allegations of corruption against one of his political advisers, which proved unfounded, contributed to Livingstone's downfall in 2008. Unlike the other interviewees, he doesn't want to see a draft of this chapter: nothing I would write about him could be any worse. I ask him if the attacks ever "get to him" and he smiles: "Not really – in many other countries I would have been killed."

CHAPTER 15

Progress in other British cities

After many years studying continental cities, when planning this book I began to pay more attention to success stories in Britain. London's story was well known and relatively well documented; others were much less so. Among the 'core cities', the eight largest English cities outside London, the amount of people who drove to work rose substantially between the 1991 and 2011 censuses. Car ownership also rose over the 20-year period, though more recently this trend has reversed in most cities.[369] Car commuting rose in Glasgow and Cardiff, although it fell slightly in Edinburgh.[370] For success stories outside London, we need to look at some smaller cities. The top 30 local authorities in England and Wales with the smallest number of car journeys for travel to work are dominated by London boroughs. However, five provincial cities are included in this group. Two of these provincial cities, Brighton and Cambridge, have made dramatic progress since 1991. The census is the only source of modal share information for most British cities although it only asks about travel to work (the future of the census was also, shamefully, under threat at the time of writing this book[371]).

Figure 15.1 shows the changes in commuting patterns over those 20 years in Brighton and Cambridge, compared to the eight English core cities[372] (excluding home working, which increased almost everywhere). The biggest changes were the rise in car driving and the fall in bus use in the core cities. In Brighton and Cambridge, cycling, walking, train and bus travel all rose at the expense of car driving.

Brighton and Hove

The origins of Brighton's transport transformation can be traced back to 1987, when the Council joined with soon-to-be-privatized Brighton and Hove Buses to launch 'Freeway', a campaign to tackle traffic congestion by reallocating road space to buses. It didn't get off to a good start: a launch event in Trafalgar Square, with Michael Palin (patron of Transport 2000), was cancelled when the 'cinema bus' used to promote the campaign broke down.[373] The original plan was abandoned in the face of local objections but the principle survived and with hindsight much of it, and more besides, was gradually implemented in the face of stiff opposition at every stage.

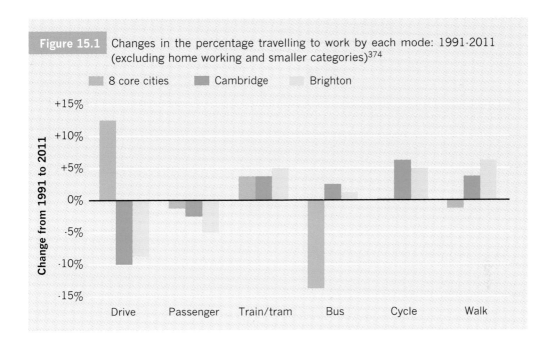

Figure 15.1 Changes in the percentage travelling to work by each mode: 1991-2011 (excluding home working and smaller categories)[374]

Brighton had never suffered the demolition and urban road building of larger British cities. Several new road schemes were proposed but never implemented. The last of them, the Preston Circus relief road, was scaled back and finally abandoned in 1991.[375] Meanwhile, to the north of the town (which became a city in 2001), a bypass was built in stages between 1988 and 1996. Gill Mitchell, who became a Labour councillor in 1993 remembers:

> "It drained a lot of the through traffic out of the city almost overnight and it was some far-sighted political thinking, mainly by Steve Bassam, the council leader, who said: 'We've got to ready the city for the future. **It's growing; it's going to continue to grow**', so looking ahead it was decided to get those bus lanes in. It was hugely controversial. **There are still some people that say we lost wards, we lost elections because of that.** Looking back it seems strange that there was such a fuss but that's what happened, and I think it forged the partnership between the bus company and the council."

At that time, the highway authority was East Sussex county council (Brighton and Hove became a unitary authority in 1997) but a joint committee of the two authorities gave Brighton and Hove councillors a say on transport issues. Despite differences of political complexion the two authorities generally cooperated well.

Figure 15.2 shows the bus lanes that were built in the central areas of Brighton in the 1990s and 2000s. These and several others were opposed by a mixture of disgruntled residents and small businesses, eagerly publicized by *The Argus*, the local newspaper. This extract from a letter published in 2002 was typical of many:

> "Buses have their use for people who are too young to drive, are unfit through physical or mental infirmity or cannot afford a car or motorcycle. They are unlikely to be the greatest wealth producers. One cannot imagine a captain of industry riding on a bus. Time is money and buses should not have priority. Bus lanes reduce road width and cause congestion and it is incredible they were even considered."[376]

Like many opponents of sustainable urban transport, the author of this letter lived in a more suburban location (Worthing). In some cases, proposals were abandoned in the face of local opposition, but several more were installed, some of them outside the central area.

The bus company played a key role in this strategy. In the run-up to deregulation the Thatcher government split the National Bus Company (NBC) into smaller units ready for privatization. The Brighton and Hove division of NBC was bought out by its management, as was the council-owned Brighton Buses. Gill Mitchell thought this was:

> "one of the best things which happened for the city, looking back. We had a group of people who knew how to run buses, who had a commitment to the town... They wanted to create a bus company that had the backing of its passengers and wanted to work in partnership with the local authority..."

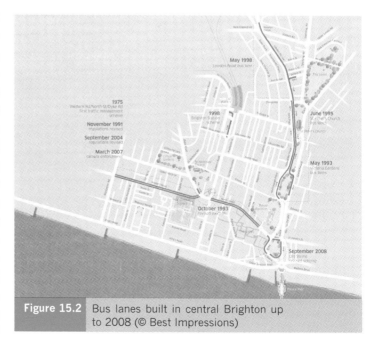

Figure 15.2 Bus lanes built in central Brighton up to 2008 (© Best Impressions)

Both management buyouts were short-lived. As in other parts of the country, the big operators, keen to build regional monopolies, were circling. In his book *Pride and Joy*, Roger French of Brighton and Hove Buses wrote:

> "By the early 1990s we had become surrounded by the now Stagecoach-owned Southdown which was a new plc group with something of an aggressive reputation to gain market share. We decided, with some personal reluctance that the best option...was to sell the company to one of these emerging groups..."[377]

Brighton and Hove Buses sold out to Go-ahead, which also acquired Brighton Buses in 1997, merging them into a semi-autonomous subsidiary, run by many of the same people. Unlike local management of the big operators in most cities, Brighton and Hove Buses was able to innovate and take a longer-term view. Roger French, who became divisional managing director, gave a public face to the company and actively engaged in local debate on transport issues. He led the Brighton and Hove Business Forum and the city's Transport Partnership, using his positions to push for greater bus priority:

> "These pro-bus highway measures have been vital ingredients, as there was no sense in our investing in increasing frequencies if the extra buses were just going to become stuck in congestion..."

A bus lane opened in 2009 along the coastal A259 from Peacehaven was one of the most controversial. The A259 was, and remains, highly congested at peak times. The bus lane was planned under a Labour administration, which lost power to the Conservatives just before it was due to be implemented. Mitchell, who was responsible for transport in the Labour group, explains:

> "The contracts had been let. There was no way the Tories could pull out of that, even though they would have liked to..."

As the scheme came under attack, French invited journalists from *The Argus* to join him on a ride along the new lanes. They organized a challenge demonstrating what many people were now finding: the bus was now quicker than the car.[378] Mitchell says:

> "It was hugely successful... people were coming to work saying: 'I got the bus today – brilliant!' They really bought into the idea of getting to work swiftly by bus..."

Peacehaven is typical south coast suburbia, the sort of place you might imagine the last bus chugging over a hill at six o'clock in the evening with three pensioners and a hefty subsidy. Catching a bus there on a Sunday afternoon in October was a surreal experience: 10 minutes after I saw the last one leave another one arrived more than half full, and the service runs until 11.35pm on Sundays. I have never been there on a weekday morning but the timetable shows a bus to Brighton every 3 to 5 minutes.

Simplified ticketing, comprehensible route diagrams and an early introduction of real-time information at bus stops were among the innovations and basic good practice introduced by Brighton and Hove Buses. Although Brighton and Hove dominated the market, other operators continued to run some long-distance and subsidized routes. As French wrote:

> "In 2010 it does seem inconceivable, yet it is true that in some parts of the country commercial bus companies are still producing timetable information without letting passengers know the detail of journeys that may be run by another company in the evening or on Sundays when they are funded by the local authority. **We have always aimed to provide a fully comprehensive picture of what is available to make things as easy as possible for our customers.**"

New visitors to Brighton will be struck by the giant pictures of local people, instead of commercial advertising on the sides of the buses (Figure 15.3). In the early 2000s the management of the company realized that much of this advertising was promoting the motor industry: they decided to phase it out, and use their buses to advertise their own service. Some of the people shown were local celebrities. A person with attitudes like the Worthing resident quoted on page 186 refused to believe that such people really were using the buses and complained to the Advertising Standards Authority. The complaint was dismissed.[379]

As part of their informal agreement with the bus company, the city council pegged parking charges to bus fares, to ensure the bus remained competitive. Increases in parking charges were never popular, and as bus fares began to outstrip inflation, the policy was eventually abandoned. A large proportion of Brighton and Hove is covered by controlled parking zones, which have been gradually extended since the late 1990s.[380] These extensions proved as controversial as the bus lanes. As long ago as 2001, there was a year-long waiting list for residents' permits in some areas. In the densest areas, the council was issuing 1.7 permits for every space reserved for residents; a council spokesman pointed out that a permit did not guarantee a parking space.[381] The leader of the Conservative group, who had campaigned against controlled parking complained that "if people are paying for a service they expect to receive that service", but the basic problem, it seems, was too many cars in a finite space.

Figure 15.3 Advertising on Brighton and Hove Buses

Figure 15.4 New England car-free quarter

The population and density of the borough were increasing rapidly. The Sussex Downs Area of Outstanding Natural Beauty (AONB) – made a National Park in 2009 – constrained the growth of the town. House prices were rising, family houses in the inner areas were being converted to flats, and new development was being built at higher densities, including some high-rise blocks. One response to this, following practice in London, was a planning policy favouring car-free housing in some locations (eg the New England Quarter in Figure 15.4). Mitchell explains:

> "As we were increasing the density, building on brownfield sites, we knew we weren't going to have the parking spaces, so that's why we introduced the car-free policy, bitterly opposed by the Conservatives... they thought it was an infringement of liberties, developers wouldn't build the flats because nobody would want to buy them if they couldn't own a car and so on, but it seems to have worked OK. They are still being built."

Partly because of these changes, car ownership, which was rising in the 1990s, began to fall after 2001. Several of the people I interviewed talked about the changing demographics as the seaside town became a city in more than just name. Chris Todd was a local campaigner for Friends of the Earth, and was co-opted on to the planning committee in the 1990s. He mentions the increase in student numbers. Many of them stay in the area after graduating.

> "The social structure was changing. **Many of these people were well educated... it was quite a rebellious city at the time.** People involved in squatting, the anti-roads movement. **There was a lot of political and environmental awareness...**"

The Green party won their first Brighton council seat in 1996, and grew in strength at subsequent elections. Ian Davey, whom I interviewed (see below), studied environmental policy at one of Brighton's two universities and was a cycle campaigner before his election as a Green councillor in 2007.

Apart from the bus lanes, the Labour administration had begun building cycle routes, along the sea front and several other places. In 2005, Brighton became one of six cycle demonstration towns, with government grants to fund promotion and new infrastructure. Davey and Todd felt some of these efforts had been "tokenistic" but cycling was increasing and becoming more political. A project to build a cycle path, by reducing the width of the road beside the council offices in Hove, was planned and approved by the Labour administration, and built under the Conservatives. Following their experience with the coastal bus route, the Conservatives changed their view on the newly built cycle lane, proposing to remove it and return the road to four lanes. This became a key election issue in 2011. Davey explains:

> "We started this campaign. It went national. The Tories said no one wanted the cycle lane; we got 3,500 people to sign a petition to say they did want it. Even people who didn't like the cycle lane thought that spending a million pounds to dig it up wasn't a good idea."

The Greens won the seat from the Conservatives and became the largest party on the council. Davey became the chair of the transport committee for a couple of years. As a minority administration, the Greens had to rely on support from one or more of the other parties. The Labour group had fallen into third place and targeted their fire on the Greens. Transport was a key issue of contention but their differences were more about process than substance. Mitchell says a Labour administration would pursue similar policies, but with more consultation. Davey complains that Labour councillors failed to support the more controversial measures. The partial pedestrianization of the Old Town, for example, was approved with support from the Conservatives.

Although Brighton's politics have been volatile, transport policy has progressed in a consistent direction: gradual urban improvements, extensions to controlled parking zones, speed reduction and reallocation of road space to buses and bikes. A 'Dutch-style' hybrid cycle path was a first for the borough (Figure 15.5), as were the 'floating bus stops' (where the cycle path diverts behind the bus stop) on the Lewes Road, which opened in 2013 (Figure 15.6).

One key difference between the Greens and the other parties relates to the park and ride system, which Brighton does not have. The bus company has long argued for park and ride, and most of the councillors agreed with them in principle, but plans had always foundered on the difficulties suggested sites raised in terms of planning.

Figure 15.5 Old Shoreham Road hybrid cycle path

Figure 15.6 Road space reallocation and a 'floating bus stop' on Lewes Road

A proposal to build a park and ride within the soon-to-be designated South Downs national park was rejected by councillors in 2005, when the council officers advised them that they might lose an expensive public inquiry.[382]

Davey believes the high use of buses from surrounding areas is partly due to the absence of park and ride. Todd adds that it may also contribute to high levels of rail use for local journeys. Research in other cities gives some support to this view. Studies in Oxford and York found that between a quarter and a third of park and ride users who made the same journey before the facility was built, used to travel by bus.[383] Nearly half of those surveyed in Oxford were making a new journey. This illustrates the point made in Chapter 5: any new transport service will generate some additional trips.

Even though the transport policies of the various political parties were fairly similar, the Greens were keen to force the pace, which gave their opponents opportunities to attack. A proposal for 20 mph zones provoked the ire of taxi drivers among others. Davey explains:

> "The GMB and Unite [trade unions] appeared to be using 20 mph as a recruiting device, to show they're the best union to represent taxi drivers. We've got some of the worst casualty rates in the country. We're trying to civilize the streets. The taxi companies tried to take us to court, but they bottled out at the last minute."

Lianne, the Greens' political adviser, added:

"We've had a lot of petitions from residents and community groups asking for 20 mph or traffic calming measures, but now that we're pressing ahead with it, the loudest voices are from those who are opposed to it. **When it's a community group, it's often people like school mums... [who can find it] very intimidating.** If you're not very involved in the political process, do you want to be the one standing up to a large group of taxi drivers and business organizations saying: 'No, I do want this'?"

Every move towards sustainable transport had been controversial, but opposition was sporadic and disorganized until 2013 when several of the usual suspects formed a group called 'Unchain the Motorist', whose website and full-page newspaper advertisements attacked every aspect of the Greens' policies on transport and several other issues. (The Advertising Standards Authority upheld a complaint against one of these advertisements, finding it to be "misleading" and lacking substantiation.[384]) Some of the attacks in the media and online have been personal. I ask Davey the same question I put to Ken Livingstone: does it ever 'get to him'?

"Yes, it has and it does get to me. **It's difficult to keep going against a barrage of vociferous opposition...** but despite all this, road safety's improving, air quality's improving, modal shift is happening and the city is thriving."

Explanations for the changes in Brighton's travel patterns

Figure 15.7 shows the travel to work by each mode in Brighton, in 2011 and 1991, with car travel falling and all other modes rising over the two decades. Are there any other explanations for this modal shift? In this section and the last one in this chapter, we are more interested in the reasons for *change* than the reasons why things were as they were in the first place. People sometimes confuse these two; so, for example, the fact that Brighton and Cambridge (like Freiburg and Groningen) have a lot of students may help to explain some of the cycling or bus use in those cities, but it does not explain why cycling or bus use *rose*.

Most of the key demographic changes in Brighton were similar to those of other cities, although its population grew nearly twice as fast as the eight core cities (a 19 per cent increase from 1991 to 2011). The number of students did increase by a quarter between 1998 and 2008 (to 14 per cent of Brighton's population) but this was actually less than the national average increase.[385] Most students do not appear in

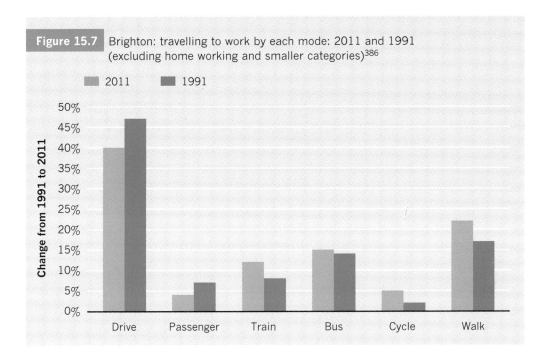

Figure 15.7 Brighton: travelling to work by each mode: 2011 and 1991 (excluding home working and smaller categories)[386]

the travel to work statistics in any case. Like most cities, the population of Brighton grew younger. The proportion of ethnic minorities increased slightly. One notable difference was the proportion of homeowners, which fell much more rapidly than elsewhere. Unlike the national average, the proportion of one-person households fell slightly, as more people shared housing. But households with children, a group who usually own more cars and drive more, rose by three percentage points.

As shown on Figure 5.4 (page 44), national bus use has fallen almost continuously since deregulation. Brighton was one of very few towns and cities to buck this trend. (Nottingham and Oxford are two other notable exceptions.) The story of municipal bus companies and NBC subsidiaries bought out by their management and sold on to the big operators happened in many other places. Most of those towns and cities followed the usual pattern of decline. Why was Brighton different? Several of the interviewees credited the leadership of Roger French. French himself believes that "deregulation has been the saviour of the industry", that the situation would have been even worse without it, but he acknowledges that things have not always run well elsewhere. He ascribes the difference to consistency of leadership. Much of the industry had suffered from a "culture of sacking managers" in the past.

The census figures for Brighton and Hove surprised several people I interviewed there. The fall in commuting by car was expected but the reputation of Brighton as a sustainable travel city rests largely on the performance of its bus company, whereas Figure 15.7 shows most of the modal shift went to walking, rail and cycling. The council's own statistics show bus patronage nearly doubled from 1992 to 2011.[387] There are several possible explanations for this apparent discrepancy.

- One point everyone agreed was the increase in bus use was not just off-peak. Brighton's buses are more frequent than they were and are pretty full in peak hours.
- The census asks about travel to work, but not to places of education: travel surveys of the two universities showed 16 per cent and 27 per cent of students travelled by bus.[388] Whether their travel patterns changed over the period is not recorded, but their numbers certainly increased.
- The census question requires people to choose one 'usual' mode of travel, the longest one by distance, so people who take a bus to a railway station would generally tick 'rail'. Rail commuting was growing, and some of those people would take a bus for part of their journey, but this would not affect the census figures.
- The council's measure of bus journeys in the borough does not differentiate between residents and non-residents. Some of the people filling Brighton's buses are travelling from surrounding areas but as Figure 15.7 is based on place of residence they do not appear there. The largest source of inward commuters to Brighton and Hove, 10,749 of them in 2011,[389] is Lewes District, which surrounds Brighton to the east and includes places like Peacehaven. The proportion of Lewes District residents commuting by bus nearly doubled in the 20 years to 2011.

When asked why commuting by train had increased, some interviewees said more people were commuting to London: high house prices were forcing many people out of London, some of whom were travelling back to work, and other local people were attracted by more jobs and higher salaries in London. Commuting to London by train grew by 92 per cent between 1991 and 2011.[390] The common assumption that people who commute from Brighton by train are nearly all heading to London is wrong, however: nearly half of them were travelling elsewhere in 2011. Rail commuting to places outside London also increased, though not quite as much (by 64 per cent) over the same 20-year period.

Cycling and walking in Brighton

Changing demographics as well as improvements to infrastructure (and possibly the marketing efforts of the Cycling Demonstration Town) will have contributed to the increase in cycle commuting shown in Figure 15.7, though it's unlikely that they are entirely responsible. The largest change in Figure 15.7 is the increase in walking to work. This was partly explained by more people living close to their place of work; the proportion of commuters with a journey of 2 km or less grew from 33 per cent in 1991 to 40 per cent in 2011.[391]

Some interviewees commented on how Brighton is a compact city which is relatively easy to walk around, but that factor hasn't changed. Traffic calming has improved the walking environment in a few areas, but this would not explain the size of the shift across the whole borough. For some other reason people were choosing to walk, cycle and take the train to work instead of driving. All of this suggests (but does not prove) that the 'push' factors – the things discouraging driving, like rising population density, less road space, rising congestion and constraints on parking – were at least as important as any of the 'pull' factors encouraging people to travel by other means.

The future for Brighton

Difficult choices on traffic and parking management are likely to become more pressing in the years to come. A planning inspector had recently rejected the draft Local Plan, instructing the council to increase its new housing allocation from 11,300 to 20,000. Mitchell sees difficult times ahead, whoever controls the council, or the national government:

> "Up to now we've had a careful planned growth policy, clustering new homes along sustainable transport corridors, but the planning inspector has asked the officers to look at all our urban fringe outside the national park. So all of our green open spaces have got to be reviewed and brought forward for development? Even cemeteries could be looked at: you could pack some houses round the edge, sports facilities – it's quite extreme! People hate tall buildings, but they are going to become increasingly a fact of life if we're to provide the new homes."

Cambridge

Pressure for house building has been particularly acute in Cambridge, a city with a booming economy and one where the fall in car ownership began much earlier, in the late 1990s. The economic boom was known as the 'Cambridge Phenomenon'; the first hi-tech companies linked to the university were launched in the 1960s, but they only began to change the city as a whole in the 1980s.

Of all the cities studied in this book, its population rose most rapidly – by a third between 1991 and 2011. In 1950, Cambridgeshire county council had published the 'Holford report', which defined the city's planning strategy for the next half century. The greenbelt, drawn very tightly around the city, remained more or less unchanged until 2006.[392] The area which had already been built upon, with its medieval centre surrounded by urban green spaces, absorbed the vast majority of the population increase but beyond the greenbelt overspill there were expanding commuter towns and villages. The first of what would be several new settlements, Cambourne (see Table 9.1, page 105) began construction in the late 1990s.

People I interviewed and many articles on Cambridge blame house prices for forcing people out and encouraging inward commuting. A couple of sources suggest the city, county and regional trends were all similar: prices more than doubled from 2000 to 2012.[393] Although there was nothing unusual about the overall trends, the price of flats grew more rapidly than houses – the average price of flats trebled over the same period – suggesting that first-time buyers may have been forced outwards.[394]

In 2013, the DfT published a report which showed that the amount of people cycling in Cambridge was increasing at a faster rate than everywhere else. In Cambridge, 47 per cent of the population cycles at least once a week. In second placed Oxford it is 25 per cent. Even more astonishing, rural south Cambridgeshire comes fourth with 22 per cent.[395] I had heard or read all the usual explanations: it's a university city, it's flat, the students aren't allowed to drive to the colleges, but the more I looked into it, the less satisfying these began to seem. Very few students are included in the travel to work figures. The terrain may help to explain why cycling was popular in the first place but it hasn't got any flatter since 1991. In places like York, with many similar features, cycling has declined but the rise in Cambridge continues. Cycling and public transport often substitute for each other but bus and train use were also rising, while car commuting fell from 42 per cent in 1991 to 34 per cent in 2011. I was missing some other element. Why was this happening?

With that thought in mind, I took a week off work, cycled across the country (a bit more slowly nowadays) and put the question to Jeremy Smith, Acting Head of Transport Policy and Strategy at Cambridgeshire County Council. His answer surprised me and triggered what would become a key conclusion to this book.

Cambridge's basic problem – and the solution

The basic problem, Smith explained, was as follows:

> "Cambridge has a mainly radial road network. **The roads that converge on the city centre are saturated.** They all clog in the peak hour. Short of knocking down listed buildings there's not much we can do about that. There were plans to build a four-lane inner ring road until the 1980s when the thinking began to change. Capacity elsewhere in the city was limited: what was the point in trying to cram more in?"

CHAPTER 15 Progress in other British cities

The alternative solution developed during the 1990s, which became known as the Cambridge Core Traffic Scheme, would restrain traffic and create filtered permeability on a bigger scale than anything attempted in any British city before or since. According to Smith, this combination of physical constraints and deliberate filtering was the key element I was missing. Everything else they did, to promote cycling, walking and the use of public transport, was supported by declining traffic and shrinking road capacity in the central areas. The road closures did not stop anyone from driving anywhere, but they made car trips across the city centre much less convenient.

Dave Earl, a campaigner with Friends of the Earth and co-founder of the Cambridge Cycling Campaign, who had led a visit of Campaign members to Groningen in 1997, described the Core Traffic Scheme as "the Dutch model". Figure 15.8 illustrates the six stages of pedestrianization and closure to general traffic.

There were several elements to the Core Scheme, including the four new park and ride sides surrounding the city and improvement to bus stops. The six closure points shown in Figure 15.8 are mainly controlled by rising bollards (eg Figure 15.9), activated by electronic passes or 'transponders', with varying levels of access for buses, taxis, emergency vehicles, and general traffic at some times of the day or week. Disabled badge holders can apply for a pass or transponder to activate some of the bollards, depending on their needs.

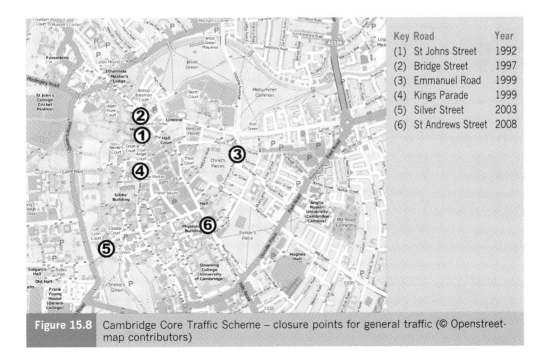

Key	Road	Year
(1)	St Johns Street	1992
(2)	Bridge Street	1997
(3)	Emmanuel Road	1999
(4)	Kings Parade	1999
(5)	Silver Street	2003
(6)	St Andrews Street	2008

Figure 15.8 Cambridge Core Traffic Scheme – closure points for general traffic (© Openstreetmap contributors)

The Bridge Street scheme was a critical turning point. The county's transport planners had devised the scheme but other council officers were unconvinced. As Earl reflected 17 years later in 2014:

> "The traffic engineers, apart from the ones devoted to cycling, still take traffic flow as their number one priority, even though that's not the council policy and never has been. They feel that if you do anything that might not clear all the traffic at each cycle of the traffic lights they've failed. So this was a major, major decision for the council."

Most highway engineers (not just in Cambridge) tend to believe that roads should only be closed if alternative routes can carry all the traffic displaced. But as Smith explained, there was no convenient alternative. There was an irregular ring of roads, but with insufficient capacity. That was the whole point: to reduce the traffic, not displace it.

Many people were unhappy with details of the scheme, particularly as it affected surrounding roads. But Earl, just back from Groningen, could see the bigger picture. In the Campaign's newsletter he defended their strategy of unequivocal support:

> "Up until May, the County Council was committed to a programme of traffic restraint. Then we had the election, and now that commitment is wavering... I think there is every chance that they will interpret expressions of only qualified support as objections... A really brave Council with lots and lots of money could put all the 'carrots' and 'sticks' in place together very quickly; but we have neither, so they have to start somewhere... in March the decision on the table will be on Bridge Street as it is now, not as a one way street instead, or closing Victoria Road, or charging motorists to deter them entering the City. On that decision – with all the implications it has for cyclists – hangs the possibility of doing all those things in the future, or letting Cambridge stew in its pollution."

The Council received over 700 cards and letters of support, outnumbering the objections by over three to one. The Bridge Street closure went ahead and with hindsight Earl believes:

> "it set the scene for future closures, making it easier for councillors to put forward the later ones. It also made the Cycling Campaign a force within the city. It surprises me how influential people think we are".

Figure 15.9 Bridge Street bus gate

Figure 15.10 Gwydir Street filtering point

The scheme provoked the usual opposition nonetheless. Addressing a conference in 2002, transport manager Bob Menzies wrote:

> "As engineers and transport planners we are often poor at public relations… the public perceived these schemes as isolated measures being implemented almost at random… we had to get the public to understand that the whole strategy added up to more than the sum of the parts…"[396]

The next stage in 1999 would coordinate the opening of the fourth park and ride site, bus priority measures and road closure number 3 shown in Figure 15.8. Menzies continued:

> "We could not convince our press office to let us use the name 'Integration Day'. They were concerned that if it all went wrong the press would quickly term it 'Dis-integration Day'. They should have had more faith."

The implementation went "without a hitch" on the day and subsequent surveys showed most of the public thought it had made the centre a better place. Pedestrian footfall in the shopping areas increased by 4 per cent. A majority of businesses and organizations based in the centre supported the scheme and appeared to understand that leaving things as they were would have made the situation gradually worse.

Figure 15.11 shows the traffic trends on the central cordon (which measures traffic crossing the river Cam in the city centre) and on radial routes around the city, compared to the trends for the county as a whole. As general traffic declined, the number

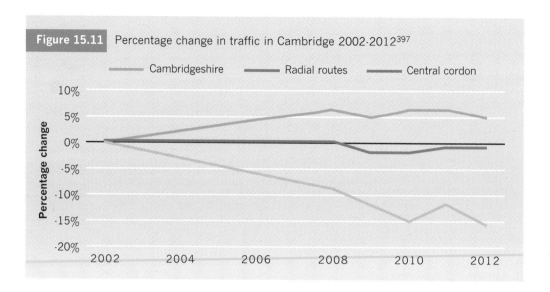

Figure 15.11 Percentage change in traffic in Cambridge 2002-2012[397]

of buses and particularly bikes increased, by a third across the central cordon, and by two-thirds on the radial routes. All of this occurred despite the rising population.

Cycling and filtered permeability in Cambridge

Cycling has always been an important mode of transport in Cambridge. I asked Dave Earl about the origins of the city's cycling culture:

> "It was not so much that cycling was promoted, it was more that it didn't lose the cycling culture that virtually everywhere had. Part of the reason was that students were not allowed, are still not allowed, to bring cars to college. Among the student community, cycling was the main way of travelling. That culture also extended to the university staff: cycling was very common among many of them. Within the inner city that culture was ingrained in the university. It's a dry area; it's flat. I don't think Cambridge had much inward migration until the 1980s. It was a small market town based mainly around the university."

Like many others, Earl moved to Cambridge to work for one of the growing number of small to medium-sized hi-tech companies.

> "Many of [the people in this sector] had roots in the university, having absorbed that cycling culture. Most of them lived in the city, or on the edge of the city, so within cycling distance: it's quite a compact city."

The science parks which grew around the city and in some nearby villages gradually spread the cycling culture over a wider area, helped by the construction of cycle routes to some of them.

Compared to most British cities, Cambridge's network of cycle routes is pretty good – nowhere near Dutch standards but moving in the right direction. Most of this progress has been made in recent years. Plans to build cycle paths along a bypass planned in the 1950s (Figure 15.12) were never implemented for some of the reasons discussed in Chapter 7 (page 65). The longstanding paths across the Commons (public green spaces since medieval times) provide useful short-cuts, but most of the older facilities are rather poor: shared pavements, narrow lanes and so on. The green spaces around the city centre provide the sort of creative walking environment described in Chapter 8 but the sharing of space between pedestrians and a growing number of cyclists has created tensions in some places, particularly where paths or shared bridges are narrow.[398]

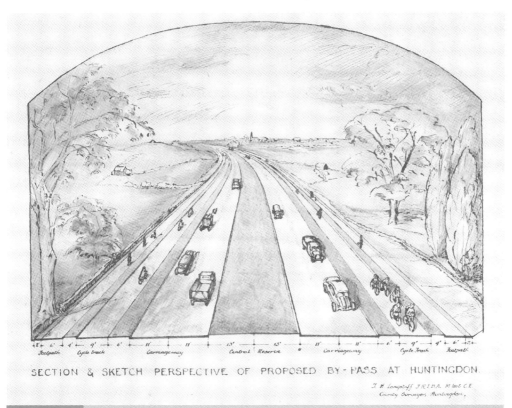

Figure 15.12 1950s plans for cycle paths alongside A14 Cambridge Bypass (Cambridge county council)

Figure 15.13 Cycle and pedestrian bridge over Cambridge Station

Figure 15.14 Guided busway and cycle path

In more recent years, prodded by the Campaign, the highway authority improved the quality of its infrastructure. The Southeast Cambridge Cycle Route is a good example: starting with an enclosed bridge over the main railway line near Cambridge Station (Figure 15.13), it combines filtered permeability through quiet residential streets and a separate cycle path leading towards the suburb of Cherry Hinton and the village of Fulbourn. The government awarded this route a National Cycling Award in 1998, following a nomination from the Cycling Campaign.[399]

Two of the best cycle routes were finished in 2011 – the paths alongside the new guided busways, which lead south to Addenbrooke's hospital and north-west to St Ives (Figure 15.14). Paths that lead outside a city are normally considered leisure facilities, but these are clearly used for transport cycling in both directions. The southern section links to several important employment sites. Many people also cycle in by bike from surrounding villages for work or other purposes.

The wide, smooth tarmac paths which are such a pleasure to ride on weren't going to be like that at first. The authority was obliged to provide a maintenance track, which could be used by cyclists, but funding for asphalt was not included. While the Cycling Campaign ramped up the pressure, Smith explains that "we spent ages scrabbling around for £2-3m out of a £120m scheme". It was transport charity Sustrans and a "rather tenuous" link to a cycling-to-schools fund which finally filled the gap.

The Core Scheme in the central areas was followed by similar moves to remove through traffic from a ring of residential areas a little further out, such as Petersfield: see the Gwydir Street filtering point (Figure 15.10). I spent a few days riding around these roads and paths, but only when assembling the figures for Table 15.1 did I realize the extent of filtered permeability across Cambridge.

CHAPTER 15 Progress in other British cities

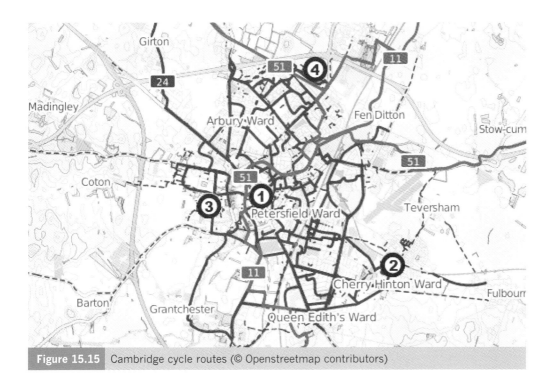

Figure 15.15 Cambridge cycle routes (© Openstreetmap contributors)

Table 15.1 used Google Maps to compare the distances between the four points shown on Figure 15.15. This is not a rigorous comparison (and it is only as good as Google's local data), but it illustrates how many journeys across the city are now shorter by bike than by car. The cycle routes also avoid most of the main roads which clog up at peak times. In 2012, the Cycling Embassy of Great Britain (page 77) visited Cambridge and was generally impressed with what they found, except for the main

Table 15.1 Filtered permeability in Cambridge: distances between four points shown in Figure 15.15

From	To	Miles by car	Miles by bike	Difference
1) City centre	2) Cherry Hinton	3.5	2.7	-23%
1) City centre	3) Robinson College	2.2	0.9	-59%
1) City centre	4) Science Park	3.5	2.9	-17%
2) Cherry Hinton	3) Robinson College	4.3	3.6	-16%
2) Cherry Hinton	4) Science Park	5.2	4.6	-12%
3) Robinson College	4) Science Park	3.6	3.6	0%
Total:		22.3	18.3	-18%

roads, which are mainly unsegregated and intimidating. Smith explains that public opposition to felling trees is often "the deal breaker" preventing cycle paths along these roads.

Public transport in Cambridge

Partly because of the high rates of cycling, travel by public transport was historically low in Cambridge. Compared to Brighton, the rail network is less useful for short to medium journeys, because there are no suburban stations within the city (although a new station by the science park on the north-east of the city is planned). Of the 3 per cent who commuted by rail in 2001, two-thirds were heading to London. The growth of employment in London helped increase this figure in the following decade.

Smith explained that the bus network was reorganized in the 1990s, with some bus lanes introduced, partly to support the new park and ride operations. There was a perception that the service was in need of improvement. Earl put it more strongly:

> "If they'd left it as it was, I think the buses would have died out altogether in Cambridge – it was such a terrible service when I first came to Cambridge. **As congestion was increasing, the buses were getting stuck in traffic.** It was a completely unreliable service, and there weren't enough of them."

Under pressure from the county council, the main operator, Stagecoach, reorganized the routes, reduced fares and increased frequencies, while the council installed some bus lanes (though not as many as in Brighton). As part of the fourth stage of the Core Traffic Scheme a formal Quality Partnership was introduced for the central area, which aimed to reduce emissions as well as improving the service. The council's monitoring report shows a steady increase in the number of buses and an improvement in their emissions standards.[400]

The biggest improvement to buses in Cambridge, which was also the most controversial, opened in August 2011, a few weeks before the census date. This was the guided busway – the 'misguided busway' as it was nicknamed by its critics. It was built along the track of the old railway line from Cambridge to Huntingdon, via St Ives and a planned new settlement, Northstowe. It uses a fairly low-tech system with guide wheels attached to the side of a conventional bus (see Figure 15.16) running on two narrow concrete paths. Because the guided buses can also run on normal roads, they can travel to towns and villages that are close to, but not on, the busway, linking them with various employment sites, the city centre and the railway station.

Many objectors would have preferred to have the rail line reopened. Others objected to the disturbance of a route which had reverted to nature since freight trains stopped using it in the 1990s. There were several reasons why the county chose a busway rather than reinstating the railway line. As rail use was growing across the region, the capacity of the final stretch into Cambridge Station would have been very limited. Widening that line would have been too expensive, according to Smith. The other reason related to flexibility. Few of the potential trips were likely to start or finish near one of the stations.

The project ran past its deadline and over budget. A legal stand-off between the county and the contractor was resolved with an out-of-court settlement in 2013.[401] Better news followed the opening of the route: it exceeded its passenger forecasts, obliging the operators to buy more buses and increase the frequencies. An evaluation study showed a higher than usual modal shift: over a quarter of commuting journeys on the guided buses and nearly half of health care journeys were previously made by car.[402] Passengers I spoke to on the guided buses were more positive about them than the critics, who continue to attack the system in the media and online.[403] The ride, in leather seats along the smooth guideway, is a pleasure, until you join traffic jams on conventional streets in Cambridge. Improving bus priority within the city is a longer-term objective for the county.

In the debate about busways versus trams, the Cambridge busway provides ammunition for both sides. Even with the legal costs, the £150m price tag was still considerably lower than a tram scheme would have cost in Britain, although many continental countries have built tramlines for less. Some advocates of bus rapid transit argue that it can match trams at lower cost.[404] That may be almost true in theory but politics prevent it in practice. A tram system would have forced the authorities to take road

Figure 15.16 Cambridgeshire busway guide wheels

Figure 15.17 Cambridgeshire guided busway – crossings

space and provide a dedicated route into the centre. A heavy rail line would have required either reinstated level crossings or – because the UK national rail authorities dislike level crossings – a combination of bridges and road closures, providing opportunities for filtered permeability. The busway designers took the politically easier option of traffic lights and gaps in the guideways, which force the buses to slow down or stop at several crossings between Cambridge and St Ives (Figure 15.17).

Political consensus in Cambridgeshire

Unlike Brighton, local government in Cambridge has remained two-tier: Cambridgeshire county council is the highway and transport authority while Cambridge City and South Cambridgeshire are planning authorities. The three have generally worked well together despite political differences and changes of political control. Earl describes the county council as "officer led", which helps to explain why in 2012 it approved, under Conservative control, a transport strategy for Cambridge including the following objectives:[405]

- Sustainable modes will be given priority over general vehicular traffic.
- Road space will be reallocated to sustainable modes at key points on the network where they suffer delay or safety issues.

Smith describes the Liberal Democrats, which is the strongest political party in the city, as "more pro-cycling" and the Conservatives as "more pro-bus", but otherwise sees little difference between the various parties on transport issues. One trend which might disturb this consensus is the growth of the UK Independence Party (UKIP) in the rural and suburban areas. An anti-cycling rant from one of its candidates was recounted in Chapter 3 (page 25). UKIP's leader in Cambridgeshire has also attacked the traffic management policies in the city, arguing that it was "already far too restrictive for cars".[406]

Bristol – and the influence of universities

Universities and students have been mentioned several times in this book. The direct influence of students is often exaggerated, but Cambridge does illustrate how universities can influence the transport culture of a city. The pattern of sustainable travel by students in Cambridge and some of the cities mentioned in this book does not apply everywhere. The main campus of the University of the West of England (UWE) in the Bristol suburbs is well served by buses, rail and reasonable cycle routes, but I found in 2011 that 57 per cent of the undergraduates in my department were travelling to lectures by car.[407] In 2006, UWE decided to employ a travel planner, Steve Ward, to improve the transport experience which many students complained about. He would also aim to rein in the growth of car travel, which would prevent the university from expanding, as the planning authority struggled with rising congestion on the surrounding roads. At first the strategy was all about 'carrots rather than sticks':

> "We need to create a really good bus service; to improve cycling; we need to make driving to university more pleasant; we need to do car sharing – it was all positive stuff. The idea was: we do need to look at car parking some day but we need to make everything else better. It's that thing you hear a lot: 'we'll only get out of our cars if x happens and y happens' and I can see it because, frankly, the bus service was terrible so you had to be seen to be making a positive impact."

The university began running its own tendered bus services. Initially the questionnaire surveys showed a shift in the right direction, but Ward was suspicious of the results:

> "We were getting increased reporting of particularly cycling, even though we looked out there and saw no more bikes on the ground… people were being lectured at about travelling more sustainably and were more likely to self-report positive behaviour, which was more in line what they thought we wanted to hear."

This problem of falsely positive responses to surveys has been noticed elsewhere.[408] The solution was to switch to cordon counts at all the entrances to the site, such that cars, bikes and bus passengers were counted and everyone arriving on foot was asked how they travelled there.

Since then, I've learned what everyone else has learned. You may make existing bus users or cyclists happier but it's only when we started to move away from minimal parking fees, with no restrictions on who could park here, that we started to see any modal shift.

While the staff unions railed against relatively modest parking charges, the decision to remove parking for new undergraduates, with a few exceptions, was implemented quite easily. A few students find other places to park, but controls are gradually tightening around the campus. Fears that it might discourage enrolment applications proved unfounded (and it hardly seems to have affected Cambridge University).

Over the following seven years, UWE's decision to start running its own bus services also had some wider consequences:

"If we look back to 2006, we had smoky 20-year-old double-deckers – everything running on a half-hour frequency, buses constantly overloaded. I had a look at a timetable from 2007: one of the services took an hour and 14 minutes to get here from the centre… [5 miles by the most direct route]. From 2007 we specified everything: this is the quality, this is the age of the vehicles, this is the driver training standards, these are the fares, these are the routes. It was almost like a London franchise model… That was very successful. We've had an annualized growth rate of 23 per cent. With that have come great pressures, such as lack of capacity at the beginning of term and an enormous admin burden managing the contract… At the same time as that, we've noticed, the commercial operators have significantly improved their offer as well. We know they've taken some ideas from us directly, and we've pushed them more than they would like to have done on things like fares, smartcards, driver training standards, frequency and so on. There was a time we were driving the competition away but the competition [First Group] has now come back with an improved offer… If that is the new reality, do we need to be running buses at all?"

Some of these changes, including significant fare reductions, have benefited the whole city, not just the services to the university. Though still some way short of Brighton's services, at the time of writing, bus use was rising across the Greater Bristol area.

In 2012, Bristol voted to have a directly elected mayor. George Ferguson, the first mayor, was an architect who had spent time in continental cities and been influenced by the best practice he observed there. He was elected on a programme which included commitments to change travel patterns across the city and improve urban public space by reducing the impact of traffic. His first major step, to extend residents' parking zones, provoked a similar public reaction to the one described in Brighton. An experimental pedestrianization once a month ('Make Sundays Special') proved much more popular. Most of the public realm changes made in 2012 to 2014 were of the usual minimalist type found in many British cities (minor widening of pavements etc). Whether Bristol will provide a British example of public realm improvements to rival any of the continental cities discussed in this book remains to be seen.

Liverpool and concluding comparisons

To conclude this and the previous chapter, it may help to contrast the three cities with another, Liverpool, whose characteristics and local policies have both been very different. Of all the major cities in Britain, Liverpool suffered the longest and deepest

CHAPTER 15 Progress in other British cities

Figure 15.18 A5036 severs central Liverpool

Figure 15.19 Merseyrail underground station

decline in its population, which nearly halved from 1931 to 2001.[409] European Objective 1 status helped to fund major road building from the 1980s onwards, including a six-lane highway that severs the waterfront from the city centre (Figure 15.18). When the recovery began in the early 2000s, it was a city with low car ownership, plenty of space and, in many places, spare road capacity.

The tone of Merseyside's local transport plans was markedly different from those of the three other cities, with economic growth "the single most important consideration".[410] This was expected, in 2006, to create "rising demand for travel". Congestion as observed elsewhere was judged to be "not yet a problem for Merseyside", although the DfT's road congestion tables show speeds on Liverpool's A roads in the morning can equal those of most of the big cities including London.[411] European funds were also used to improve public transport – Merseyrail (Figure 15.19) is one of the better city regional rail networks in the country – and some efforts were made to encourage walking and cycling.

Figure 15.20 compares the changes which took place in Liverpool with London, Cambridge and Brighton between 2001 and 2011. Although its increase in household income was the lowest of the four, car ownership rose substantially (from a lower base), while it fell in the other three. The expected increase in Liverpool's car traffic did not occur, although the fall was less pronounced than in the other three. A greater difference can be seen in commuting behaviour: the factors dissuading people from driving to work in London, Brighton and Cambridge clearly did not apply to Liverpool.

The second part of this book set out to describe who did what in the hope that patterns will emerge to help explain why certain changes occurred in some cities and not in others. Although the picture is complex, some patterns have emerged to inform the conclusions in the next chapter.

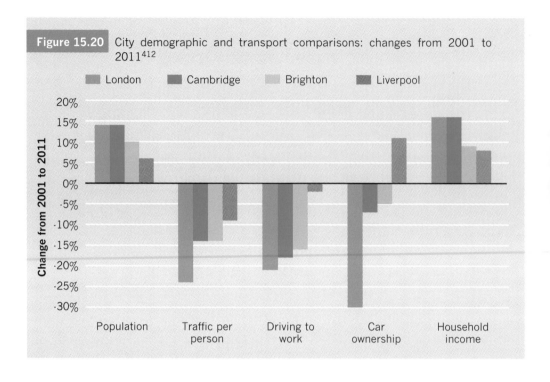

Figure 15.20 City demographic and transport comparisons: changes from 2001 to 2011[412]

CHAPTER 16

What sort of cities do we went?

This book has reviewed a wide range of evidence, on myths and problems and from cities which have made inspiring progress. As argued in Chapter 4, evidence alone can never tell us what we ought to do. In making the leap to conclusions and recommendations, this chapter will interpret the evidence from the starting point I set out in Chapter 1: it is, in other words, a personal view. Though it stresses the advantages of each course of action, we should bear in mind that all the positive changes described in Chapters 12 to 15 created political challenges and unintended consequences; there are no ideal solutions. In deciding on the cities we want in the future we should be guided by hopes and aspirations but also by the risks and problems we are, or are not, willing to put up with. Climate change, early deaths from air pollution and public space dominated by traffic would be top of my list of things we should not be willing to put up with. Presumably most of us would also like to see cities with much less traffic congestion but how many of us would accept the measures needed to make that happen?

Congestion will always be with us

Whenever politicians and business leaders mention transport, the word 'congestion' usually follows. Road building, rail expansion, bus rapid transit (BRT) and road pricing are all portrayed as weapons in the fight against the evil of congestion; and yet congestion never seems to get much better. After a brief improvement during the recession of 2008 to 2012, average speeds on the roads of Britain's core cities have continued their long-term downward trend.[413] (The 'peak car' fall in driving *per person* has been offset by more people and more cars.) There is a wide measure of consensus about the cause of the problem if not the solution. Providing there are enough people with enough vehicles wanting to use a road, vehicles will continue to fill it until congestion slows everyone down and some people start looking for alternatives, like travelling off-peak, using a different mode or avoiding some journeys altogether. There are, broadly speaking, four potential solutions:

1. Squeeze more traffic into a finite road space.
2. Rearrange the road and the area around it to accommodate more traffic.
3. Persuade people to reduce vehicle movements, or travel at different times.
4. Restrain vehicular access.

Under Solution 1, limited improvements can be made by traffic management measures. Highway authorities have worked hard with computer programs and control centres to manage traffic in the most effective way. Some further improvements may be possible in future, but the gains are likely to be marginal for as long as drivers make their own decisions on when and where to travel and at what speeds.

Transport planners and highway engineers may believe they are making big inroads into congestion because they see traffic running more smoothly after they have made changes, like expanding the capacity of a junction, for example; but that impression is misleading. Expanding capacity at one point in a road network has two effects: in the short term, it allows traffic to run more freely to many other points where congestion will get slightly worse; in the longer term, it allows the total volume of traffic to increase. So lots of projects easing congestion at junctions will not produce the same effect at the city-wide level. Some people associated with the shared space movement have made the same mistake in calling for a 'mass switch-off' of traffic lights.[414] Just because switching traffic lights off reduces queuing at a particular junction does not mean that spreading the idea will reduce congestion across the whole city. Many cities in developing countries have far fewer signal controls at junctions but that has not enabled them to avoid urban congestion.

As discussed in Chapter 11, expanding the capacity of roads or junctions cannot solve congestion on its own; the main factor generating traffic and congestion is the intensity of activities – homes and businesses – in any defined area. To make a difference in existing cities, Solution 2 would mean 'knocking down and spreading out'. This would disperse people and businesses over a wider area. It would increase total traffic and disperse it over a wider area, which might reduce congestion but would increase carbon emissions and the loss of greenfield land.[415] It would also reduce the intensity of economic activities in our city centres at a time when cities are competing to do the opposite. In Britain today, with a high and rising population density there is already resistance to greenfield building for *additional* housing. Now imagine that, on top of that, a politician announces plans to knock down homes to build new roads and reduce the number of people living in your area (a scenario which frequently occurred in the 1960s and 1970s): would you vote for him?

Solution 3, voluntary behaviour change, can help in some circumstances but as discussed in Chapter 5 positive measures, such as public transport improvements, are not a very effective way of reducing traffic on their own. They are more likely to generate more travel and even where they persuade some drivers to leave the car behind, this frees up space for others to drive. Freeing road space for more people who want to travel may be a worthwhile change, but it will make no difference to congestion.

Under Solution 4, some of the restraining measures discussed in this book would reduce traffic volumes but not necessarily congestion. This would be true of pedestrianization, road closures and reductions in road capacity. They may be managed in a way that avoids making congestion any worse but they are unlikely to make it much better.

Many economists and transport academics from widely differing political perspectives believe the solution is clear: road pricing with higher prices on more congested roads and at more congested times. Clearly this solution could solve the problem providing the authorities were free to vary the prices with no constraints. How high the price would need to rise to reduce congestion would vary from place to place and from time to time. In central London, a charge of £11.50 per day (increased by 130 per cent over 11 years) – plus billions of pounds invested in public transport – has been just enough to stop congestion getting any worse, but not enough to improve it. Average traffic speeds in inner London (wider than the congestion charge zone), which reached a dizzy peak of 13 mph during the recession, have since resumed the same downward trend of other cities.[413] An extension to the congestion charge zone introduced by Ken Livingstone was reversed by his successor, with considerable public support.

Road pricing is not the only financial tool available to authorities wanting to restrain traffic and congestion. Petrol taxes and parking charges have both been widely used over many years. There were clear signs at the end of the 1990s that rising fuel taxes were restraining overall traffic levels. Then came the Fuel Tax Revolt of 2000, since when fuel taxes have been declining year-on-year with the support of all the main political parties. Similar forces have constrained the ability of local authorities to use parking charges as a traffic management tool. UK households spend an average of just 90 pence a week on parking charges[416] but authorities that raise them are often punished by the voters. Government ministers have also tried to make political capital by curbing the ability of local authorities to enforce parking restrictions.[417]

So national road pricing could solve the problem of urban congestion if the prices were set high enough but that is not likely to be a vote-winner. Even if a government were to introduce such a scheme, its successors would come under intense pressure from voters to reduce the charges. A more likely scenario, which was approved as policy by the Liberal Democrat party in 2013, is 'revenue-neutral' road pricing, where the money raised by road charges is offset by an equivalent cut in other taxes on driving.[418] Some of the balancing cuts could come from vehicle excise duty (an annual tax on car use) but most of them would probably come from fuel duty (which is a much bigger revenue-raiser). This raises a further problem: if congestion charging was offset by cuts in fuel taxes, it would make driving more expensive on roads which are full, and cheaper on roads with room for more cars. The overall effect would be more driving, more carbon emissions and only limited improvements to congestion.[419]

So have the integrated policies of cities like Freiburg, Groningen and Lyon – combining Solutions 3 and 4 – solved the problem of urban congestion? Comparable statistics on congestion are not available although the evidence of modal shift and traffic counts (in the inner-city areas) suggest that congestion would have been worse under a 'business as usual' approach. That said, all three cities still experience traffic jams at peak times, as do Brighton, Cambridge and London.

Like some other writers on this subject, I have come to the conclusion that as long as people are free to own, park and drive cars and politicians are influenced by public opinion, urban congestion will always be with us (how autonomous vehicles might change the situation is discussed below). For over 50 years since Buchanan wrote his report, governments and transport planners have directed most of their efforts towards a problem that cannot be solved; imagine what could have been achieved if all that energy and money had been directed at solvable problems. If we set congestion to one side, the list of those other problems is long and some of them – like climate change, community severance and early deaths due to air pollution – are, in my opinion, far more important. Reviewing the evidence on congestion, American economist Anthony Downs gave the following advice to commuters stuck in traffic jams:

> "Get a comfortable air-conditioned vehicle with a stereo system, a tape deck and CD player, a hands-free telephone, perhaps even a microwave oven... Learn to make congestion part of your everyday leisure time, because it is going to be your commuting companion for the foreseeable future."[420]

While I agree with his assessment, I believe we can do better than that. If we recognize that urban congestion cannot be solved, we can devise better strategies for dealing with it: mitigating its worst effects and offering people alternatives to sitting in traffic jams where they are unavoidable.

It's not the economy, stupid!

Former US President Bill Clinton reportedly kept a sign over his desk during his first election campaign, saying: "It's the economy, stupid!" The same sign hangs metaphorically over the desks of transport professionals and politicians concerned with transport almost everywhere. The fixation on alleviating congestion has been paralleled by the elevation of economic considerations. The two often go together; for example, the Confederation of British Industry (CBI) declares (with no reference to any evidence) that "congestion costs the UK economy up to £8bn a year" in a report arguing for more road building.[421] The Coalition government upped this claim to £19bn a year when announcing its return to large-scale road building in 2013; this claim was also made without reference to evidence and without explaining why they believe road building would improve congestion or the economy.[422] Chapter 4 briefly reviewed the evidence on the links between transport and economic activity. Transport investment can encourage firms and economic activity to move from one place to another, though not always in the expected direction. A new road may attract companies to towns along it, or it may encourage them to move away. Whether transport investment makes any significant difference to national *economies* has never been proven.

An academic review of the research evidence on this question concluded:

> "Public sector investments in transportation infrastructure result in long-term economic benefits ... A preponderance of the studies reviewed could not reject that hypothesis."[423]

Having read many of the same papers, I don't recall many of them seriously trying to reject that hypothesis. Transport academics have a vested interest in this issue for two reasons. In the current climate, if academics can show that certain types of transport changes bring economic benefits, politicians are more likely to listen to them. Second, decisions about research funding tend to follow the same logic. To admit that transport has only a limited economic impact could be seen as an argument against funding transport research. So at the risk of arguing myself out of a job, the evidence reviewed in Chapter 4 casts doubt on many of the claims made about transport investment and the economy.

Clearly a completely dysfunctional transport system could have a negative impact on economic activity, as in developing countries that do not have a national network of sealed roads, but the economies of cities in developed countries have proved remarkably resilient to road congestion. A recent American study found that richer cities, on average, have more traffic congestion than poorer ones.[424] London is the most congested city in Britain and it is also the one with the most productive economy.[425] The strategy of Transport for London has all been about providing alternatives to driving – alternatives that serve people travelling for business or any other purpose. Instead of the Bill Clinton slogan, a more appropriate sign to hang over desks at the Department for Transport and the transport departments of local authorities might read:

> "If the transport strategy works, the economy will look after itself."

What we mean by works may vary from place to place, but three common factors I have found in all the successful cities I have visited are:

1. integration of cycling, walking and public transport;
2. modal shift towards these three;
3. improving the quality of urban life by restraining traffic and creating more public space.

To improve transport and our cities we must save green fields

Whether those three principles can be achieved in a densely populated country like Britain will depend on how we respond to rising population and the growing need for housing. Three broad strategies are possible:

1. knock down and spread out;
2. fill in and spread out;
3. intensify and minimize greenfield development.

As discussed above, the first of these three would be highly damaging to our environment and way of life. Fortunately it would also be costly and politically impossible for any government to implement on a wide scale.

Most planners and national politicians seem committed to Strategy 2, coupled with the shift away from flats and towards 'family housing' (particularly houses with four bedrooms or more) described in Chapter 9. Though not as profligate as 'knock down and spread out', Strategy 2 will still require a lot of greenfield land.

The projections show the UK population growing by over a third between now and 2087 – an extra 23 million people.[426] Even if that is an overestimate, population growth will cause greater congestion whichever strategy is chosen. Although none of the strategies will solve the problem, each one will have different implications for the nature and distribution of congestion. Under Strategy 2, the residents of new car-based suburbs will add to existing traffic moving to or from towns and cities and the motorway network, particularly at peak times. The conventional response of UK planners that we can avoid this problem by designing new suburbs around 'high-quality public transport' (ie bus routes) is wishful thinking for the reasons discussed in Chapters 5, 6 and 9. Restraining car use is difficult, if not impossible, in new suburbs; the only exception to that rule is where suburbs have dense urban centres, like the district centres of outer London, which have frequent rail services and controlled parking. Where parking and car ownership are unconstrained, the vast majority of travel is always done by car. Look at the roads leading in and out of your nearest big town or city at peak times and imagine them with thousands more additional cars. That (plus large-scale greenfield building somewhere near you) is where 'business as usual' and most of our politicians are currently leading us.

The pressures are most acute in southern England but all parts of the UK will experience population growth. The claim that we could solve the problem by building more in the less affluent north is exaggerated. A more active regional policy, redistributing employment towards northern cities would be a good idea for many reasons, but it would make no difference to the greenfield land requirement unless it was part of Strategy 3, intensification, which would be my preferred solution.

All of the cities described in Chapters 12, 14 and 15 apart from Liverpool[427] had intensified to varying extents, either as a deliberate policy or because of constraints such as the London greenbelt and the South Downs national park around Brighton. Intensification was an important element in the relative success of all six cities. The pressures to accommodate more people forced politicians to make, and voters to accept, difficult choices like parking controls and reallocation of road space to buses and bikes, which they might otherwise have deferred or avoided. They forced the UK national government to grant London the powers and funding to address the transport crisis it faced in the early 2000s. If London had been allowed to sprawl like many American cities, it would now cover much of south-east England and its patterns of transport would be very different. At a national level, the intensification policies pursued in the UK between 2000 and 2011 helped to stem a rising tide of traffic across the country that had lasted for over 50 years; similar policies and changes also occurred in several other countries.

The mention of urban intensification, of raising population densities, often invokes negative images of high-rise concrete blocks and people living in rabbit hutches. In reality, most of the beautiful places where people live in cities around the world are high in density: think of central Paris, Amsterdam or Bath. A renewed programme of intensification, planned in the right way, could provide homes for Britain's growing population, restrain the growth in traffic that would otherwise occur and save large swathes of rural Britain from sprawling development. It could also help to improve affordability in the housing market – a problem which the last wave of intensification did not address because of its concentration on high-value city centre sites.

A complete answer to Britain's housing problems would fill several more books: I can only look at part of the problem and part of the solution here. There is widespread agreement that Britain's high housing prices are socially divisive (and possibly economically damaging). There are many things governments could do to bring down housing prices, particularly through tax changes, but no government is likely to make any radical changes in that direction, for two reasons: it would create too many powerful losers and it would depress the economy in the short term. If those rational policies are politically impossible, the next best solution would expand supply of housing at the lower end of the market ie smaller flats and terraced housing (England could also follow Scotland's example, to stop selling off and start expanding its social housing stock).

As most high-value central sites have now been redeveloped, a major programme of urban intensification would need to find new sites in inner-city districts, suburban centres and even some sites at the edge of cities where good access to public transport exists, or can be provided. These locations will provide less expensive housing for first-time buyers and an increase in supply will also drive down rents. Of course, some might prefer a detached house with a big garden, but what we prefer is always constrained by what we can afford. At the moment, the first step on the housing ladder is too high for many legs to reach. Current policy, which encourages construction of large 'family houses', will not create any more families, nor will it address the problem of affordability; it is more likely to create a nation of reluctant house-sharers.

Giving intensifying cities the 'wow factor'

If intensification is to be reinstated as a guiding principle, deciding where and how will have major implications for transport. Figure 9.12 (page 104) showed the relationship between settlement size and traffic. Larger towns and cities generate less traffic than smaller towns and villages; London is in a category of its own. Within these categories, inner areas and denser neighbourhoods generate the least traffic. Put simply, if we want to minimize overall traffic, we should concentrate as much new development as we can in bigger cities, and constrain it most in rural areas. Intensification always entails some shift towards more sustainable transport, even where supporting changes are minimal. Where the demand for parking exceeds available space, parking has to be controlled whether voters or politicians like it or not. Demand for public transport increases with population density making it easier to provide more frequent services. Local shops and other services become more viable and people walk more as a result. All of these effects, but particularly the parking controls, tend to reduce car ownership. These changes reduce traffic generation, but on their own they are not enough to prevent local traffic worsening, and with it the quality of life in intensified areas (remember the paradox of intensification from Chapter 9). When asked what they don't like about their neighbourhoods, residents of inner urban areas consistently rank traffic and related problems at the top of the list.[428] To improve the attractiveness and quality of life of intensifying cities, traffic has to be removed. Lyon, Cambridge, Hackney and particularly Groningen show how this can be done: closing roads to through traffic, pedestrianizing and reclaiming public space.

London's transport planners have discovered how incremental improvements like pedestrian crossings, wider pavements, bus and cycle lanes can improve the local environment and reduce motor traffic in intensified areas. Installing a crossing to improve pedestrian safety is an easier plan to sell than installing one to squeeze traffic out of an area. Removal of parking always provokes opposition but, like pedestrianization, it offers the prospect of a win-win solution, freeing space for sustainable modes and reducing the attractiveness of urban driving. An increase in commercial car parks to offset any losses may be a price worth paying: companies hoping to profit from them could prove useful allies in the fight for more sustainable transport.

Some of these measures, though necessary, will be perceived as negative restraint imposed by unsympathetic authorities. To succeed politically, any strategy of urban intensification must provide positive, visible improvements – the type of change that everyone who sees will feel: "Wow – what an improvement! (and I certainly wouldn't want to go back to how it was)." The banks of the Rhône in Lyon, Freiburg's Altstadt and Groningen's Grote Market are all good examples. The planners of British cities have been much more timid by comparison, too ready to accept second-rate compromises. The growing influence of the shared space movement offers politicians a way to create the appearance of change without really changing anything. Exhibition Road in London (Figure 8.1, page 80) is one example, where £29m was spent on largely cosmetic changes. If civic leaders are to carry their voters and their taxpayers with them on this journey, they will need to be bolder and more imaginative than this.

The quality of the environment from a pedestrian viewpoint is vitally important for intensifying cities, not just to encourage modal shift, but because it is one of the key elements that make some cities attractive places to live, work and do business in. As discussed in Chapter 8, green spaces, pedestrianized areas and joined-up walking routes with minimal interruptions are all factors that promote creative thinking and problem-solving – important to cities that rely on knowledge and creative industries. I have no idea what difference this makes to GDP but these principles have been understood for many years in old university cities like Cambridge; Michel Woitrin also understood it when planning Louvain-la-Neuve.

Car-free development can provide traffic-free walking routes for residents and the wider area. A national programme of urban intensification will multiply opportunities for car-free development. Where parking is controlled, road capacity is limited and planners are trying to fit more housing into a limited space, its advantages are obvious. Car-free residential developments could also breathe life into many of our existing pedestrianized areas. Private property developers have embraced the limited form of 'car-free housing' in inner London. To go further, to build a Vauban or a Stellwerk 60, takes leadership from local politicians and planners.

As I was putting the finishing touches to this book, I was contacted by transport planners at TfL, gathering advice and ideas. Following the Roads Task Force, they were looking at longer-term trends with some concern. The speed of population increase could unravel all the gains of recent years, unless car ownership and use come down at a similar rate. They were looking at a range of options, including the European model of carfree development. Whatever emerges from that review, London will provide a useful example over the next few decades; an extreme case from which other cities may learn (in some respects, although continental cities may still be more inspiring in others).

One new city is much more sustainable than ten new towns

Urban intensification should be the main element of any strategy to accommodate rising population but the potential for intensification is constrained by the availability of suitable sites and the economics of housing development. We may also question whether the UK can sustainably support 86 million people – a figure that may be reached within the lifetime of younger readers – and how such growth might be averted. (This raises issues outside the scope of this book; we may simply observe that population growth is not easy to restrain in a democracy with an open economy.[430]) But whatever policies are pursued, some greenfield building will be inevitable.

There is an emerging consensus among mainstream politicians in favour of small new settlements as one possible response. For the reasons explained in Chapter 9, small new settlements are a particularly unsustainable form of development however they are dressed up. They are also an irrelevant distraction from the urgent need to provide more housing in the UK. The garden cities prospectus specifies a minimum of 15,000 dwellings for a new garden city but even that is much larger than most developers or local authorities would contemplate in practice. The examples listed in Table 9.1 (page 105) are all much smaller than that. Of the current generation of new settlements under construction, Northstowe in Cambridgeshire has a target of 10,000 dwellings[431] and Cranbrook in Devon just 5,000.[432] After all the arguments around the eco-towns programme just 4 small urban extensions were built, but let us consider a 'best case' scenario where 10 new towns with an average population of 40,000 each could be privately developed with the approval of local authorities by 2035. Those towns would house just 4.8 per cent of the additional population projected between now and then.[433]

If some greenfield building is unavoidable, new cities (with a target population of 200,000-250,000, like Brighton, Freiburg and Milton Keynes), planned around the principles contained in the eco-towns guidance and the recommendations in this chapter, would be a much more sustainable option. Accompanied by intensification and a tightening of greenbelts around other towns and cities, this policy could save far more of our countryside than any other option; it would also produce the most sustainable patterns of transport.

But there is a catch: no local authority, or local voters, will ever agree to such a plan and agreements between private developers and landowners will never produce anything on such a scale. To plan new cities would need something like the New Towns Act, which remains on the statute book for any government brave enough to use it. Most of the development can be done by the private sector, providing the land is purchased by the government, which also finances the infrastructure. That aspect of the original New Towns programme worked well; less successful aspects such as the low densities, ugly concrete structures and car-dominated transport planning all provide useful learning experiences. With the benefit of experience, we could build new cities as populous as Milton Keynes but smaller in size, producing more sustainable, better places to live. Although it cannot be achieved by consent, this strategy also offers political advantages. Government ministers in the late 2000s were reportedly surprised by the negative reaction to the eco-towns programme. They shouldn't have been; a cleverer strategy would have sought to limit the amount of people they were likely to upset with plans for new settlements.

Public transport to support intensification

Some elements of the public will always oppose any form of restraint on driving or parking, even when there is no alternative (ie there are too many cars in an area that cannot be expanded). Other people may accept restraint with varying degrees of reluctance providing they are accompanied by significant, visible improvements in public transport. Any strategy for intensification and urban improvement needs to address that political imperative. New tram systems have proved a powerful political tool in French cities and to some extent in those British cities that have built them, even if their direct contribution to modal shift is limited. Whether bus rapid transit systems (BRT) can help to win that argument is less clear. The Cambridge system, which is arguably the best example in Britain, remained controversial after its implementation and a new BRT network planned in Bristol (offering fewer advantages than the Cambridge system) was proving very unpopular in the planning stages.[434] Rail improvements are, by contrast, welcomed by almost everyone (providing traffic to and from stations is not allowed to increase).

To support a policy of urban intensification we need public transport systems that enable more city dwellers to live without owning a car. Car ownership is falling in our cities and among younger people at the moment, so much of what follows would apply regardless of any change in planning policy. A renewed policy of intensification would accelerate the fall in the cities and by concentrating more people there would also increase the national car-free population. This would have important implications for public transport planning. Whereas cycling and buses tend to substitute for each other for shorter journeys, for longer-distance travel people without cars need access to rail.

Many local authorities and some private transport consultancies conduct what is known as 'accessibility planning'. This measures journey times by public transport and/or walking to key destinations. These are usually regular destinations within a short distance such as employment areas, shops and health services. For people who choose to live without a car, access to longer-distance and less regular destinations, eg family members in other parts of the country, is also essential.[435] In the inner areas of larger cities, where most of these people live, there is nearly always a mainline station somewhere nearby. Although some will hire or borrow cars or join car clubs for occasional journeys, if they are planning to move somewhere without rail, they are more likely to buy a car. One obvious implication is that rail is vital for intensifying areas. If we accept that urban congestion is unlikely to disappear in the foreseeable future (and may well get worse), then rail expansion is a far more effective response than putting more buses on to congested roads.

Chapter 9 discussed some of the myths around travel to work. Although leisure and shopping generate more trips, the ability to get to work affects car ownership and where people choose to live. Intensification and declining car ownership in cities will necessitate what planners and transport planners should be doing anyway. Public transport (and cycle) networks need to connect any point where someone lives to any place of work across a city-region. The design of transport networks and the location of jobs both matter. The challenge is easier to address in intensifying cities or new cities than in sprawling suburbs or small new settlements.

Connectivity is partly about frequencies of buses, trams or trains, partly about how they connect with each other and partly about how close they go to each end of a person's journey. The mode also makes a difference. Most people are more willing to change to or from a train than to change buses.[436] Few people would choose a two-bus journey to work or any other regular destination if they could drive instead. All of this suggests that urban rail improvements, including some new lines and stations, need to accompany intensification.

The traditional approach to transport planning assumes the importance of a transport route is related to volume of demand: destinations with the highest volumes are the most important and should be given the most resources. This thinking has led to High Speed 2 – an expensive plan to expand capacity and reduce journey times along corridors which already have the best inter-city rail services in the country. Local and regional connections to the planned high-speed stations leave much to be desired. For a growing urban car-free population, rail connections to many destinations will be more important than high speed on one section of one or two journeys.

There are also implications for rail pricing. Pricing is now used to remove most potential travellers from many peak-time journeys. Peak travel into London will always be scarce and rationed, but penal pricing also applies on lightly used routes such as the hourly Cross Country services between the north and the west country. A standard return from Bristol to Newcastle, for example, cost £303 in 2014 – three times the equivalent fare for the slightly longer but quicker journey from Frankfurt to Hamburg. Fares like this suggest much suppressed demand. A shift from driving to rail for longer-distance travel would require both expansion of peak capacity and reductions in peak fares.

London and Freiburg both suggest that increased subsidy combined with comprehensive easy-to-use ticketing can boost demand, enabling the subsidies to be substantially reduced a few years later. A policy of this nature would be impossible where buses are deregulated and the authorities are legally prevented from competing with commercial services. The current arrangements seem to be working well in places like Brighton but not in most British cities. Re-regulation offers no panacea. However, where current arrangements are clearly not working, local authorities need the option to re-regulate and to increase subsidies in the medium term in order to improve value for money in the long term.

In 2014, the Scottish government was consulting on strengthening the powers of local transport authorities to regulate bus services in their area – much to the chagrin of the big bus companies, two of which are headquartered in Scotland. As described in Chapter 5, the UK government created several powers for English local authorities to re-regulate buses, but it then sat back and allowed the big bus companies to make legal threats against the few authorities that considered using these powers. This subversion of public policy by vested interests is a national disgrace; the UK government (and the devolved nations if necessary) should legislate to protect local authorities in that situation and should remind the bus operators where most of their revenue comes from and who they are answerable to – the taxpayer.

I have not said much so far about rural transport. Figure 9.12 (page 104) illustrates what anyone who has lived in rural areas will have experienced, that life in the country involves a lot of driving. Although car ownership has risen in rural areas, 9 per cent of rural households remain without a car.[437] Many of these people, who can be found in some of the remotest areas, are old; and if retiring to the country remains popular, their number could begin to rise again as the population ages.

A growing urban car-free population also has implications for rural transport policy which is rarely recognized. Public transport planning in rural areas concentrates on short-distance journeys from villages to the nearest towns. Connections to the rail network for longer journeys may happen by accident but are rarely planned. Even some towns in counties like Cornwall take two uncoordinated buses with walking in between to reach the nearest railway station. A growing car-free urban population will need to visit rural areas, not just for leisure but to visit family members who may have stopped driving. Better coordinated bus services can offer one solution; new technologies may offer others. Taking a taxi or hiring a car to make the final leg of a journey from the nearest station to a remote village is an expensive option. Providing the option of smaller cheaper electric vehicles available to hire from railway stations could help to plug that gap in the longer term.

Car clubs may also be considered a form of public transport; their membership has been growing in many cities as the proportion of people choosing not to own a car has risen. Demand for their services is greater where parking is constrained. Most car club members do not own a car; they mainly travel by sustainable modes, using the car club vehicles occasionally, when necessary. A smaller proportion of households with two adults use a car club instead of buying a second car.[438] Car clubs make more efficient use of limited road space in cities and they could be more effectively supported by public authorities. They work best in those areas where parking is controlled and constrained; their success depends on local authorities and/or developers allocating sufficient parking spaces in the right places.[439] Demand for their services has been particularly high in London, but at the moment car club operators have to separately negotiate with the 32 London boroughs, each of which has a different set of rules; some boroughs allocate free parking bays, while others make substantial charges. All of this has constrained the growth of car clubs in London; a London-wide approach, implemented by TfL would be much more effective.

To cycle like the dutch, learn from the dutch

As discussed in Chapter 7, attitudes to cycling have been changing in British cities; some transport planners and political leaders have been on study visits to the Netherlands and I have heard tales of Damascene conversions on their return, although some British interpretations of 'Dutch style' facilities (like the one in Figure 7.17, page 78) still leave much to be desired. TfL is moving ahead with its 'Mini Holland' conversions of district centres in parts of outer London and with its segregated east-west and north-south cycling Crossrail routes; how well these will be implemented in practice remains to be seen but TfL has at least recognized that it needs to learn new techniques.

Of all the lessons we could learn from continental cycling cities, the following are particularly clear:

- Get the infrastructure right and behaviour change will follow (as in Seville).
- Traffic restraint is a vital element.
- Cycle routes should:
  - separate cyclists from heavy traffic and traffic congestion;
  - separate cyclists from pedestrians;
  - join up in logical routes, which may combine cycle paths and quiet roads;
  - give the bike a shorter journey than the car (filtered permeability);
  - have no barriers and cater for all types of bikes and mobility scooter;
  - follow consistent design standards with no compromises;
  - minimize interruptions and give priority over other traffic.

A comparison may be drawn with the motorway network. The last two points, which also apply to motorways, help to explain why they have been so successful in encouraging car use, while cycle routes in British cities – where those points do not generally apply – have done very little to encourage cycling. Because cycle routes are considered on a purely local project-by-project basis, public consultations often sabotage their coherence in ways that would never be allowed to affect the strategic road network.

With the exception of Seville, the cycling networks in the continental cycling cities described in Chapters 7 and 12 took several decades to achieve and the same would apply to British cities; the principles above describe the direction of change that city leaders should follow if they are serious about sustainable transport. To make real progress towards them we need stronger national guidance (scrapping the hierarchy in Figure 7.14, page 75) and local authorities need to empower their cycling officers. These are usually fairly junior employees too easily over ruled by 'old guard' engineers pursuing a very different agenda. The professional bodies representing highway engineers are beginning to take cycling infrastructure more seriously and some universities (including my own) are helping to train a new generation of transport engineers with more understanding of cycling issues, but progress is slow.

Hostility against cyclists among certain sections of the population is linked to broader social tensions as discussed in Chapters 3 and 7. Local politicians and transport planners cannot resolve those tensions on their own but framing cycling issues in a different way could help to alleviate them. The Hackney approach towards public realm improvements that also benefit cyclists is one way. Another way is to widen the target beneficiaries of cycling infrastructure. Dutch cycle routes are frequently used by families with children in a range of trailers and cargo bikes and also by people with disabilities using mobility scooters and handcycles. Designing for all these people and showing them on publicity materials would help dilute the perception of public money being spent on a narrow 'out-group' of fit middle-aged men.[440]

Hostility to cyclists in the UK has been exacerbated by forced sharing of space between cyclists and pedestrians. Genuine Dutch-style separate infrastructure (like Figure 7.4, page 66, and unlike the 'Dutch-style' path in Figure 7.17, page 78) could remove that source of conflict. For local politicians struggling to improve conditions against a barrage of criticism, my advice would be: go for separation of cyclists and pedestrians as a principle and a slogan. If it is done properly (nearly) everyone can agree on that.

Forcing the pace on electric vehicles

Although it would be wrong to view electric vehicles as a panacea, electrification is an indispensable element of any strategy for rapid carbon reductions. (Another book in this series, *Sustainable Energy without the hot air*,[441] explains why electric power is a better option than hydrogen power for most purposes.) Electric vehicles can also alleviate some of the other problems described in this book, particularly noise and the concentration of air pollution, which is already at dangerous levels in many of our cities, and could be exacerbated by population increase. As many of our urban roads will remain congested for the foreseeable future, traffic jams composed of electric vehicles will cause much less harm: to the neighbourhoods they are crawling through, to pedestrians and cyclists and to drivers and passengers who breathe the same air.

Chapter 2 was critical of the Climate Change Committee's optimistic assumptions about the uptake of electric cars. Their earlier reports tended to portray a world where changes would flow smoothly from government actions, a world with few unintended consequences, although some of their more recent reports have pointed to policy gaps and areas where progress isn't happening fast enough: the take-up of electric vehicles is one of them.

Like any new technology, electric vehicles raise many technological challenges that require more research and development. Companies from different sectors are working on the biggest challenge – how to increase battery capacity – but only motor manufacturers can address some of the other challenges. Most of the larger manufacturers are still putting more research and development effort into petrol and diesel cars; they still view electric cars as a niche product.[442] A timetable towards phasing out petrol and diesel cars would provide the certainty manufacturers need to direct their research and development towards the technologies of the future – providing they understand that lobbying cannot overturn the timetable.

Another barrier to the take-up of electric cars is the proliferation of incompatible charging systems. As with many other aspects of UK transport policy, the pattern of charging points is fragmented, with several competing networks and local initiatives providing points that are compatible with some vehicles and not with others. The preference of UK governments for distributing money to local authorities through ad hoc programmes and competitive bidding has exacerbated the problem; the 'Plugged in Places' scheme, which could have helped to establish a national standard has had the opposite effect.[443] In 2014, the European Parliament adopted measures aimed at imposing some form of standardization, though its application will depend on action by national governments.

Electrification of the vehicle fleets of national and local government and other public bodies could help to lift the market for electric vehicles beyond the point where it gains its own momentum. A programme to provide electric vehicles to car clubs, car hire companies and taxi operators could also help that process. A more serious attack on air pollution when it reaches dangerous levels (as it often does in British cities) would help to encourage the shift to electric vehicles. Some cities, such as Athens and Paris, have used emergency measures to address dangerous spikes in air pollution, banning cars with odd or even registration numbers from circulating on those days. Whatever restraining measures are taken, clearly electric vehicles should be exempt.

A joined-up transport system needs joined-up decision-making

Many of the problems described in Part I stem from short-term project-by-project decision-making. Most of the solutions described in Chapters 12 to 15 required longer-term strategic planning. Transport decision-making in France, Germany and the Netherlands is more strongly guided by strategic plans. British transport authorities are supposed to formulate a strategy as part of their local transport plans but these documents generally combine broad principles, descriptions of measures already under way and a wish list of projects for the Department for Transport to fund, many of which are never implemented. The leaders of continental cities (and regions) have much greater control over their own finances – as does London, compared to other British cities. Politicians of all parties seem to agree that devolving more financial autonomy to cities would be a good thing but progress has been very slow.

Part of the problem relates to governance – many UK cities have no single authority responsible for the whole conurbation or city-region. Local Enterprise Partnerships, which can fulfil that role if they and the government agree, are undemocratic bodies; business representatives on their boards raise serious questions about accountability and vested interests. If many of them were given serious powers over transport or planning, these weaknesses would become more apparent.

The solution is clear, widely recognized and normal practice in many countries: create elected bodies responsible for transport and planning across city-regions and give them control over tax revenues raised in their area. The French transport tax has proved particularly useful in providing a long-term revenue stream to transport authorities in cities like Lyon. It is a model which the UK, and other countries, could learn from and adapt, if not necessarily copy. The transport tax is a tax on employment – one of many taxes that have made employing people expensive in France, contributing to high rates of unemployment there. The transport tax is more readily accepted by French employers than some of the other taxes because it brings visible improvements in the cities where they operate, but there may be other ways of raising revenue from businesses which would have less impact on employment.

A workplace parking levy is one option which already exists in the UK; other options might include a tax on business energy consumption. Although several cities have considered a parking levy, only Nottingham has implemented one so far. Any business tax which is set locally will leave individual councils vulnerable to lobbying and threats from businesses to move somewhere else unless the rates are reduced. To make a difference in the longer term, any tax dedicated to transport would need to be set nationally, although local rates could vary according to city size, wealth or other objective factors.

Another factor contributing to short-termism and unsustainable decision-making in British transport planning is the requirement for project-by-project cost–benefit analysis (CBA). Although CBA is used in other countries, less weight is attached to it and greater weight is attached to compatibility with longer-term strategies. If national government insists on CBA, then local decision-makers with a longer-term vision may have to find creative ways of working around it. To take one example that I know has occurred in the past, closing a road will increase the cost element of a CBA because it will increase some journey times. So if a local authority wants to close a road as part of a major project assessed by the Department for Transport, it may omit that element from the project bid, and close the road at a later stage – when it can do so without conducting a CBA.

The dumbing down of political debate associated with the 'sound bite revolution' has also worked against joined-up decision-making in transport, as in many other areas.[444] Our media and political culture discourage politicians from acknowledging that some problems cannot be solved and that some gains require sacrifices, so perhaps we the voters need to do this for them. If we vote for the promise of simple painless solutions, we are equally to blame for the failure that will inevitably follow. So if any politicians seeking your vote claim they can tackle urban congestion, regenerate the economy through transport investment, build more 'family housing' and protect the countryside, maybe you could lend them a copy of this book!

Looking to the future

Any reflections on the longer term depend on what happens to the climate. If emissions continue to rise and, as seems increasingly likely, tipping points are passed, transport systems like everything else in our civilization will be at risk. The literature on adaptation to climate change often makes the same mistake as the Climate Change Committee, assuming gradual and predictable change, when the reality may be very different. This book started by talking about climate change but its main focus, transport in urban areas, is not the main culprit when it comes to carbon emissions. The biggest contribution comes from trips between 10 and 25 miles.[445] These usually begin in a town or city but finish somewhere else. Unlike urban transport, our European neighbours offer no shining examples of sustainable inter urban travel. The analysis in this book suggests one solution which doesn't make much difference – high-speed rail – one factor which makes the situation worse – motorway building – and several other possibilities.

Electrification is probably the change with the greatest potential to reduce carbon emissions over the medium term. Urban intensification can also help; it has two effects on longer-distance travel. It replaces some long journeys with shorter ones; where people are able to walk to a supermarket, for example, they are less likely to drive to an out of town store (see Figure 9.3, page 95). But the critical factor is car ownership. Figure 6.3 (page 52) illustrates the dramatic effect of car ownership on distance as well as mode of travel. Put simply, people without cars rarely drive and do not travel as far. This remains true even when factors such as household income are taken into account. If the fall in car ownership, which has already begun in our cities, can be encouraged to follow the London path, this will make a substantial difference across the country.

In the longer term (probably longer than their enthusiasts imagine), autonomous vehicles could blur the distinction between public and private transport. If programmable electric vehicles are moving past your home and you aren't going to be driving anyway, why do you need to own a vehicle? Fully autonomous road vehicles could significantly reduce the problem of urban congestion, particularly if decisions about routing were controlled by a central computer system (unless and until it breaks down). If that sounds too much like 'Big Brother' for motorists who want to retain the freedom to drive where and when they like, then we should not expect autonomous vehicles to miraculously solve urban road congestion. In some towns or cities, systems like Ultra at Heathrow Airport (Figure 11.7, page 120) could provide new forms of public transport, although finding the space (and the money) for the network of separate tracks may prove impossible in many places.

Autonomous vehicles do not need as much road space because they can move in straighter lines. If their movements are coordinated in some way, they could increase road capacity. Those two changes could offer further possibilities: reclaiming space for pedestrians, cyclists or new forms of public transport or simply allowing more vehicles to circulate. Reflecting on these options, a recent report into autonomous vehicles for TfL posed the key question: "What type of city do we want to live in?"[446] The same question could apply to many of the more immediate issues discussed in this book. Do we want to live in cities dominated by traffic and vehicles, or should we concentrate on making them better places to live and work? And we may broaden the question: what type of country do we want to live in? Do we want to maintain the distinction between town and country or should the two be allowed to merge as new housing spreads? If you have ever considered such broad questions, you may also have wondered what you, as an individual, could do about any of them; the final chapter will suggest a few possible answers to that.

CHAPTER 17
What can I do?

In 1972, Doris Lessing published a short story, *Report on the threatened city*, which seems strangely prophetic to anyone who has tried to act on climate change, or any major environmental threat. A group of aliens arrive on Earth and take on human forms to try to warn the inhabitants of a city about to be destroyed. Some people accept the warning with fatalism and carry on as before. Some young ones take drugs, write songs about it and are treated for paranoia in a mental hospital. The aliens learn that some humans are taken more seriously than others, so then they take the forms of scientists and start presenting at conferences. But they find, to their astonishment, the participants:

> "are soothed and relieved by stating a problem, but having done this, seldom have the energy left to act on their verbal formulations".[447]

Nearly 20 years ago, I was at a seminar where someone raised a question which is heard more often nowadays: "Isn't it all too late?" The presenter's response has stayed with me as a guiding principle ever since: certainty only arrives in retrospect, but even if it is too late to avoid catastrophic change, and even if I cannot stop it, "How would I want to have behaved?"

This book has described many problems and barriers to change. Even political leaders and senior officials find it hard to make radical changes to transport patterns. If you are neither a politician nor an official, you may be wondering: what can I do? This chapter considers that question from three perspectives: local campaigning, the foot soldiers of transport planning and from personal experience of trying to 'walk the talk' in everyday life.

Local Campaigning

My own experience of leading environmental campaigns began in the late 1990s with a campaign against plans for widespread greenfield building in south Devon. This led on to campaigning for sustainable transport, as described later. The links between urban containment and sustainable transport have been widely researched but I was not aware before writing this book that it would lead me back to where I began: if we want to solve our urban transport problems, we need to constrain the sprawl of development into rural areas.

With the earlier chapters written, I went to see Neil Sinden, Director of Policy and Campaigns at the Campaign to Protect Rural England (CPRE). CPRE has a large supporter base with local branches and affiliated organizations, including many parish councils. It also has branches in some cities, including London. Changes in national planning policy under the Coalition government, which weakened the powers of local planning authorities, had led to "an upsurge in local campaigning". As the national office focuses on backbench MPs (more receptive than ministers and party leaders) CPRE encourages its branches to put pressure on councils to push through local plans and then stick to them.

As we talk about local campaigning, Sinden mentions a successful example which takes me back to Brighton. One factor that pushed Brighton and Hove councillors towards more sustainable transport was the South Downs national park, which prevents the south coast towns from sprawling inland. Campaigning by CPRE from the 1920s onwards led to two government inquiries into the concept of national parks. The wartime 'Dower Report' recommended the creation of 12 national parks including one in the South Downs. The 'Hobhouse Report' of 1947 excluded the South Downs, because expansion of agriculture during the Second World War had eroded the landscape and wildness (a key criterion in creating a national park).

In the 1960s, two Areas of Outstanding Natural Beauty (AONB) were created in Sussex and Hampshire, extending some protection to the Downs, but not as much as a national park. One of the people who led the campaign to create a national park was Chris Todd, representing Friends of the Earth:

> "Ironically, a lot of credit must go to Lord Steve Bassam. As [Labour] leader of Brighton borough council in 1995, he proposed selling off the Downland estate, helping to create a huge campaign, with a local network called Keep Our Downs Public. [The proposal] was abandoned in the face of a rebellion by a few Labour councillors and a local party very unhappy with his proposals."

This about-turn gave the campaigners an opportunity to raise the issue of a national park.

> "With the support of the local Labour parties and then promoting the issue nationally we got it on to the national political agenda. We were being helped with examples of other damage to the Downs, which the

> Countryside Commission tried to downplay… **We were lucky in the timing of some of these examples.** An area of chalk grassland restoration, paid for by public purse, was ploughed up four weeks before the 1997 general election."

A succession of MPs and parliamentary candidates visited the area, putting the issue on the agenda of the incoming Labour government. In 1999, Deputy Prime Minister John Prescott announced two new national parks, covering the New Forest and the South Downs. Christopher Napier was the chairman of Hampshire CPRE:

> "We were all delighted about that, but then began a consultation process about the boundaries. **Eventually the Countryside Agency came up with their proposals for the boundaries.** It had to go to a public inquiry if any local authority objected. **After a bit of campaign work most of the authorities were persuaded to come out in favour.** Unfortunately West Sussex county council and Chichester district council decided they didn't like the look of it at all."

Following a public inquiry where only objectors were allowed to appear in person, an inspector recommended a smaller national park, excluding key settlements such as Lewes. So began a major campaign, involving 160 organizations, which each paid a subscription to employ Todd three days a week. Membership was restricted to organizations rather than individuals but most of the work was done by volunteers. The first aim was to persuade the secretary of state to reopen the inquiry. The actor Brian Blessed and author Bill Bryson, President of CPRE at the time, helped raise the public profile of the campaign, which collected 10,000 signatures on a petition and 1,000 names on postcards. When the second inquiry was announced, a small group of volunteers prepared to give evidence. This time, they were treated as objectors, allowed to appear in person, alongside Natural England, which was also objecting. Their arguments won the day: nearly all the areas they sought were included in the National Park from 2010.

I ask Napier what lessons he would draw for successful campaigning:

> "Good leadership, good organization, volunteers with the time to commit. **A public face like Bill Bryson – someone well known.** Taking a professional approach: we never overstated our position. **Everything we said was evidenced – if you overstate your position you open yourself to attack.** And getting public support: we did a lot of

Brian Blessed and Bill Bryson supporting the campaign for a South Downs National Park[448]

presentations, to parish councils and so on, trying to get rid of the misconceptions."

Membership of a national organization, Sinden adds, can strengthen action at the local level. For Napier, and the CPRE branch, the work has continued. Outside the national park, the new planning rules and pressure from developers were leading to more housing at lower densities sprawling over the "ordinary countryside".

In my own campaigning over the years, I have found that NIMBYism can be a powerful force for better, for worse, or a mixture of both depending on the issue at stake. In 1998, I set up an organization called SHARD (South Hams Against Rural Destruction), which campaigned with some success against the large scale of greenfield building proposed in the Devon County Structure Plan at the time. The threat of big developments in specific areas helped us to mobilize large, angry public meetings, which pushed the politicians towards urban intensification in and around Plymouth (the main city in the area). Like most such campaigns, the mass reaction we managed to mobilize on those few occasions was short-lived. The effort of sustaining a local campaign rapidly falls on very few shoulders, but a big reaction at an early stage helps establish its credentials in the eyes of politicians and the media.

Another powerful weapon, we discovered, is a mole in the system. Several times I was phoned by individuals – council officers and even people working for private developers – who disagreed with the actions of their employers. One man, I remember, would always call from a public phone box after checking he hadn't been followed. A mole inside the district council told us about a planned development of 900 houses on the edge of Ivybridge, which we passed to the media, provoking outrage in the council chamber and a fruitless internal inquiry. The furore provoked by the leak strengthened the hand of the councillors for Ivybridge, who persuaded the council leadership to shelve the development.

As we heard in Brighton, when politicians respond to campaigns for positive but controversial change, the supporters often go quiet, leaving the headlines to opponents. Many people I have met through environmental campaigns are motivated by a new idea but rapidly move on to something newer, never seeing anything through to a successful conclusion. Some take on too much and 'burn out'. For others, networking becomes an end-in-itself, rather than a means to an end. Anyone publicly identified with a campaign will receive many requests to go to meetings, join committees and get involved in other issues. To remain effective and keep some sort of private life intact, I have learned, most such requests must be politely declined.

Political independence is essential for any campaign, unless it is conducted openly by a political party, but equally, campaigns and concerned citizens should be as ready to support politicians making positive changes as they are to condemn negative ones. And, of course, no changes will ever be perfect, or free from side effects. As Dave Earl argued to the members of Cambridge Cycling Campaign about the Core Traffic Scheme (Chapter 15, page 198), effective campaigners sometimes have to support things they may not like; insisting on perfection is an argument for no change.

With the trends described in the last chapter, many towns and cities will need these sorts of measures if they are to avoid being swamped by traffic as population rises. As we saw in Cambridge, removing through traffic does not prevent anyone from driving anywhere they want to (a common misunderstanding) though it does tip the balance between different modes for some journeys. Could your town or city benefit from such a change? Does a similar campaign already exist? If so, why not join it? If not, why not start one? You may find, as I found in Bristol (see www.livingheart.org.uk), many organizations are willing to support such a campaign. All it needs is one determined individual to take the initiative.

Transport planners – 'off the record'

Many people who choose transport or planning as a career – those who make a conscious choice – are often motivated by environmental values. I often wonder, and occasionally hear, how the students I have taught get on in professional life. Through a letter to *Local Transport Today*, I invited transport planners with such values, who may have disagreed with their employers, to talk 'off the record'. Those who replied were all older, when speaking out no longer threatens dire consequences. One of them mentioned a colleague who had seen the letter but was too afraid to respond.

All of them had varied careers, moving between local government and private consultancy, sometimes working in their own time for pressure groups or on local campaigns, occasionally provoking the ire of senior managers. As a result one had been selected for redundancy and another initially dismissed, but reprieved on appeal.

Several of them described dysfunctional bureaucracy in local government. Private consultancies were less hierarchical – "more efficient but not always for the right aims", as one put it. Where the client wants you to say "black is white", many consultants "believe you do what they ask", as another put it.

All agreed that senior managers in either sector rarely attach much importance to sustainability. Widening pay differentials and organizational politics tend to promote ruthless self-serving people; altruistic principles can weigh you down on 'the greasy pole'. In local government, highway engineers committed to increasing road capacity still tend to dominate, making life difficult for planners promoting sustainable modes. But all of the interviewees described things they had achieved, mainly small-scale. One consultant had found ways of giving the clients what they wanted but adding suggestions for more sustainable solutions. I asked another how he would advise a newly qualified transport planner with principles. In local government, he said:

> "Keep quiet about what you do. I found the way to do things was: don't go through your boss. **I used to get things done by making links with people in other sections.** Find the people who are good… if you do it as a cross-disciplinary thing you get a much better design… you can get access to more money and it works… if you can get away with it…"

Walking the talk

Some of the transport planners I interviewed contrasted their personal travel behaviour with senior managers, who remained attached to the car as status symbol. In 2013, I was on a panel at a conference debating the question: "Do climate change researchers have a responsibility to lead sustainable lives?"[449] A similar question could be put to campaigners. The simple answer is obviously "Yes", but the "How" is not so clear. I argued that researchers or anyone else advocating social change should try to behave in a consistent way, particularly within their areas of interest. As mine were transport and planning, I had given much thought to both.

I once had a long discussion with a climate change denier, an intelligent man who claimed to follow scientific rationalism. There was little meeting of minds between us, except when I mentioned that I didn't drive or fly. Reflecting on this for a moment he said: "Well I suppose if you really believed humans were changing the climate, you wouldn't, would you?"

I said in Chapter 8 that my experience of urban living has been a positive one : it was a radical move from a Dartmoor village to a city centre flat. As I wrote in *AtoB* magazine shortly afterwards:

"I read up about planning and housing and gradually realised that I had been part of the problem. **Three of the greatest threats: urbanisation, road building and climate change, are all linked.** Pressure for road building is both a cause and effect of more cars and more driving. **Pressure for urbanisation comes from smaller households, people living longer, and particularly in attractive rural areas, inward migration – of people like me.** An inconvenient truth gradually dawned on me: if you really love the countryside, three of the best things you can do for it are: move to a town or city, give up the car, and if there are just one or two of you, move to a flat…

Now you may be reading this, **doubting whether anyone would make such life changes for purely altruistic reasons**: surely we had other, more selfish reasons. And that's where I've made another discovery. The psychologists call it 'cognitive dissonance', a fancy term for something most of us would recognise in people around us. Put simply, **where people's behaviour conflicts with their attitudes, one or the other is likely to change.** The young radical who gets a promotion and becomes more conservative, the SUV driver who convinces himself that climate change is a conspiracy invented by governments: these are the results of cognitive dissonance – **people's attitudes changing to reflect their behaviour.**

Cognitive dissonance affects all of us at different times in different ways. In my case, I began to resent having to drive 7 miles to the cinema or 18 miles to the theatre in Plymouth. **I started to find driving more stressful. I wished I could walk to cinemas and theatres.** Although I never stopped loving the Devon countryside, the attractions of city centre life grew stronger. So when the time came to look for a new job, I was happy to move."[450]

When we sold the car I assumed we would join a car club, or hire cars occasionally, as many of our carless friends do. But in five years I have never needed to drive. I have met several people who tell a similar story. Many of the things they imagined they would need a car for, they find they don't. Online and mobile tools have helped a lot. I walked all 630 miles of the Southwest coast path, a few days at a time, without needing a car. Some of the journeys in Cornwall were convoluted but everywhere could be reached with a little planning. Along the way, I've discovered a silver lining to some of the problems described in this book. Most people never venture more than a few hundred yards from the nearest car park. I have spent whole mornings walking beautiful stretches of coast with only peregrines, choughs or seals for company.

Giving up the car is, of course, much easier if you live near a centre where buses, trains and even cycle paths, in our case, converge. The further you live away from such centres the more difficult it becomes, although wherever you live, it is always possible to make some changes to the way you travel: *Cutting Your Car Use* by Anna Semlyen contains some useful suggestions.

My wife is a community nurse. She changed jobs and a condition of her new job was that she must have a car, so she bought one. However, the car usually stays at her place of work, while she commutes by public transport. This prompted me to investigate how NHS trusts treat the travel of their mobile staff.[451] Some trusts in London and Bristol have taken a more progressive approach, encouraging alternatives and providing pool cars for those who need them. There is a curious social convention operating here. Most of us need equipment of some kind for our work: computers, printers, photocopiers, sphygmomanometers if you are a doctor or nurse: how many employers oblige their staff to buy these things themselves? Why then are cars treated differently? The answer partly relates to question number 3 in the introduction: most people imagine that owning a car is 'normal', that nearly everyone has one. The idea that such requirements will make people buy more cars than they want or need may not occur to many employers. In most cases (my wife being a rare exception) these people will then start using them for much of their private travel, exacerbating the problems described in this book. If you are an employer, you may be able to help change this situation.

In 2005, I decided to stop flying, at least for personal travel. Since then, I have had to fly once for work when there was no other alternative, but I and some of my colleagues have made a stand against the globetrotting conference culture which permeates academia. If that damages my career prospects, so be it. I was once invited to Australia, all expenses paid, to address a conference about (you couldn't make this up) 'reducing energy consumption from transport'. I presented by video link and wrote an article for the university magazine about it instead.

As we saw in Chapter 1, business flying only accounts for a relatively small part of the carbon problem. Most flying is for holidays or visiting people. Family in far-flung places may create some obligations to fly, but anyone can decide to stop flying for holidays. As with driving, once you stop doing it, you find yourself making other choices: the journey becomes part of the holiday. A serious move to more sustainable travel would entail a change in social attitudes to holidays: shorter distances for short breaks and a return to something like the Grand European Tour of the early 20th century, where people[452] take longer breaks occasionally in a lifetime and they become once again a life experience.

Epilogue

I am writing this final paragraph in Paris, on the steps that straddle the recently pedestrianized bank of the river Seine. (I remember the network of roads along the two banks as a racetrack when I lived here – the one which would seal the fate of Princess Diana.) Boats glide quietly opposite the Jardin des Tuileries where the trees are in blossom. It is a damp working day but several people are sitting around me while others stroll, cycle or skate along the old road below. Some photography students set up their tripods, while a man takes an impromptu video of four girls dancing in a line. The place still has a temporary feel. Over time, the old road surface will be replaced and it may grow to resemble the banks of the Rhône in Lyon – a monument to the folly of past generations and of hope for the cities of the future.

Endnotes

Note on web references
Many of the sources below are freely available on the internet. Only the home pages are listed, as the specific URLs are often long and tend to change. You may find them easier to find by typing the title between quotation marks into a search engine.

1 The myths of urban transport
1. From the 2011 census. Coincidentally, the figure is the same for Greater Manchester and the (smaller) City of Manchester.

2 The problem
2. There are many sources of information on these threats. See for example: IPCC (2001) *Third assessment report – synthesis report.* On: www.ipcc.ch
3. Royal Society (2010) *Climate change: a summary of the science.* London. On: http://royalsociety.org
4. DfT (2011) *Aviation greenhouse gas emissions.* Report: Factsheet 5. On: www.gov.uk:
5. DfT (2013) *UK aviation forecasts.* Appendix E3. London: Department for Transport.
6. Committee on Climate Change (2013) *Meeting carbon budgets – fifth report to parliament.* On: www.theccc.org.uk
7. Committee on Climate Change (2010) *The fourth carbon budget – reducing emissions through the 2020s.* On: www.theccc.org.uk
8. Committee on Climate Change (2013) *Meeting carbon budgets – 2013 progress report to parliament.* On: www.theccc.org.uk
9. DECC (2014) *Digest of United Kingdom energy statistics.* London: Department of Energy and Climate Change.
10. DECC (2012) *Gas generation strategy.* Report: 8407. London: Department of Energy and Climate Change.
11. See: MacKay, D. J. C. (2009) *Sustainable Energy without the hot air.* Cambridge: UIT. Also available free on: www.withouthotair.com
12. See: www.nextgreencar.com
13. King, J. (2007) *The King review of low-carbon cars.* London: HM Treasury.
14. Carson, I. (2004) 'Perpetual motion – survey'. *The Economist.* 4 September.
15. Scrosati, B., Hassoun, J. and Sun, Y. (2011) 'Lithium-ion batteries: a look into the future'. *Energy & Environmental Science.* 4 (9), pp. 3287-95.
16. Bates, J. and Leibling, D. (2012) *Spaced out: perspectives on parking policy.* RAC Foundation. On: www.racfoundation.org
17. Davis, A. and Jones, L.J., 1996. 'Children in the urban environment: an issue for the new public health agenda'. *Health & Place.* 2 (2), pp. 107-13.
18. Frank, L. D., Andresen, M. A. and Schmid, T. L. (2004) 'Obesity relationships with community design, physical activity, and time spent in cars'. *American Journal of Preventive Medicine.* 27 (2), pp. 87-96.
19. Hart, J. and Parkhurst, G. (2011) 'Driven to excess: impacts of motor vehicles on the quality of life of residents of three streets in Bristol UK'. *World Transport Policy & Practice.* 17 (2), pp. 12-30.
20. YouGov (2011) 'Attitudes to immigration: report for migration watch'. On: www.migrationwatchuk.org.uk
21. ONS (2013) *National population projections 2012 – based extra variants report.* On: www.ons.gov.uk
22. Technically two recessions in the UK: the first and deeper one in 2008-9 and a shallower 'double dip recession' in 2011-12. See: *Quarterly national accounts* on www.ons.gov.uk
23. CLG (2013) *Rents, lettings and tenancies: numbers of households on local authorities' housing waiting lists.* Table 600. On: www.gov.uk
24. Pannell, Bob (2012) 'Maturing attitudes to home-ownership'. *Council of Mortgage Lenders.* On: www.cml.org.uk
25. Marshall, Ben (2012) 'We are all NIMBYS now? Localism and development'. *Ipsos Mori Social Research Institute.* On: www.slideshare.net
26. Stratton, Allegra (2012) 'Open land can solve housing shortage, says minister'. *Newsnight.* 28 November. On: www.bbc.co.uk
27. See www.emptyhomes.com and: Gallent, Nick, Mace, Alan and Tewdwr-Jones, Mark (2004) 'Second homes: a new framework for policy'. *The Town Planning Review.* 75 (3). 287-308.

3 "There has been a war on the motorist"

28 DfT (2007) 'The NATA refresh: reviewing the new approach to appraisal'. London: DfT. This document appears to have disappeared from the national archives. The phrase was used verbally at the time, and was attacked by the shadow transport minister in a conference speech: Villiers, Theresa (2009) *'Smarter choices and low carbon transport'*. On: www.conservatives.com

29 The quote was reportedly made at a conference in 1997. It was repeated in parliament to Prescott who agreed that he had made the commitment (Hansard (1998) 20 October). This did not prevent him from trying to deny that he had ever said it, five years later: Hall, Sarah (2002) 'Prescott denies making car pledge'. *Guardian*, 6 June.

30 DfT (2013) *Transport Statistics Great Britain* Table ENV0105 (TSGB0305). Adjusted by the Q2 Retail Price Index from: www.ons.gov.uk

31 Martin, Daniel (2012) 'We're the fuel tax capital of Europe'. *Daily Mail*. 28 February.

32 House of Commons Transport Committee (2009) 'Taxes and Charges on Road Users'. Sixth Report of Session 2008/9.

33 Eurostat (2013) 'Total taxation share in the end consumer price for euro super 95 and diesel oil'. On: www.eurostat.eu

34 Hammond, Philip (2010) 'Government announces plans for next phase of high-speed rail'. Speech to the Conservative Party Conference. On: www.conservatives.com

35 HM Treasury (2013) *Investing in Britain's future*. London. On: www.gov.uk

36 I explained this point on BBC Radio Bristol recently, following a statement from a councillor who claimed that a proposed budget of £16 per person per year on cycling initiatives in Bristol was much too high for a 'marginal' mode of transport. Although much more than most British cities spend on cycling, this actually represents less than 5 per cent of public spending per person on transport across the UK.

37 HM Treasury (2013) *Public sector expenditure by function, sub-function and economic category tables*. Chapter 5, Table 5.2. On: www.gov.uk

38 There are many sources to back up this statement. See for example: Wolmar, C. (2005) *On the Wrong Line: how ideology and incompetence wrecked Britain's railways*. Rev. and updated ed. London: Aurum. And: Bowman, A., Folkman, P., Froud, J., Johal, S., Law, J., Leaver, A., Moran, M. and Williams, K. (2013) *The great train robbery: the economic and political consequences of rail privatisation*: Centre for Research on Socio-Cultural Change. On: www.cresc.ac.uk

39 Figures downloaded from the Office for Rail Regulation (www.rail-reg.gov.uk) adjusted for inflation show an increase of 2.8 times between 2000-1 and 2003-4.

40 Butcher, L. and Keep, M. (2012) *Buses grants and subsidies*. House of Commons Library. Report: SN1522. On: www.parliament.uk.

41 DfT (2012) 'Transport statistics Great Britain'. Table TSGB0123 – Historic.

42 See for example: *Taxpayers' Alliance* (2011) 'Excessive taxes on motorists in each council area in the UK'. Research Note 95. On: www.taxpayersalliance.com

43 Quoted in: Gutfreund, Owen D. (2004) *20th Century Sprawl: Highways and the reshaping of the American landscape*. New York: Oxford University Press. Page 34.

44 The comments were made by Peter Burkinshaw (2013) in response to questions from the Cambridge Cycling Campaign. 30 April. On: www.camcycle.org.uk

45 I Pay Road Tax (2011) 'Government minister sticks to his mistaken claim that motorists pay for roads'. The minister was Mike Penning, junior minister at the Department for Transport. On: http://ipayroadtax.com

46 Sustainable Development Commission (2011) *Fairness in a car dependent society*. Available, for the time being, on: www.sd-commission.org.uk

47 European Commission (2012) 'Decision of 25.6.2012 on the notification by the United Kingdom of Great Britain and Northern Ireland of a postponement of the deadline for attaining the limit values for NO_2 in 24 air quality zones'. On: http://ec.europa.eu

48 Rice, Dennis and Francis, Wayne (2006) 'Undercover probe reveals the "buckets of money" made from speed cameras'. *Daily Mail*. 15 October.

49 Massey, Ray (2013) 'Speed cameras "increase risk of serious or fatal crashes": new RAC investigation raises doubts over their usefulness'. *Daily Mail*. 7 June.

50 Allsop, R. (2013) *Guidance on the use of speed camera transparency data*. London: RAC Foundation. On: www.racfoundation.org

51 Allsop, R. (2010) *The effectiveness of speed cameras*. London: RAC Foundation. On: www.racfoundation.org

52 DfT (2012) *Reported road casualties in Great Britain: 2011 annual report.* 'Contributory factors to reported road accidents'. On: www.gov.uk
53 Rosén, E., Stigson, H. and Sander, U. (2011) 'Literature review of pedestrian fatality risk as a function of car impact speed'. *Accident Analysis & Prevention.* 43 (1), pp. 25-33.
54 Grundy, C., Steinbach, R., Edwards, P., Green, J., Armstrong, B. and Wilkinson, P. (2009) 'Effect of 20 mph traffic speed zones on road injuries in London, 1986-2006: controlled interrupted time series analysis'. *British Medical Journal Online First.* 339.
55 Christmas, S., Britain, G., Buttress, S., Newman, C. and Hutchins, R. (2010) *Cycling, safety and sharing the road: qualitative research with cyclists and other road users.* Department for Transport. On: www.gov.uk
56 Marques, J., Abrams, D., Paez, D. and Martinez-Taboada, C. (1998) 'The role of categorization and in-group norms in judgments of groups and their members'. *Journal of Personality and Social Psychology.* 75 (4), pp. 976-88.
57 Christmas, S., Britain, G., Buttress, S., Newman, C. and Hutchins, R. (2010) *Cycling, safety and sharing the road: qualitative research with cyclists and other road users.* Department for Transport. On: www.gov.uk
58 Johnson, Daniel (2013) 'Horse-rider fined for taking pony into McDonald's after being refused drive-through service'. *Daily Telegraph.* 22 July.
59 Several examples on www.bigbrotherwatch.com
60 Richards, D.C. (2010) *Relationship between speed and risk of fatal injury: pedestrians and car occupants.* London: Department for Transport.
61 DfT (2013) *Free flow vehicle speed statistics: Great Britain 2012.* London: Department for Transport. On: www.gov.uk
62 Heffer, Simon (2004) 'Why the police love to hate the motorist'. *Daily Mail.* 27 April.
63 Hogg, M.A. and Abrams, D. (1988) *Social identifications: a social psychology of intergroup relations and group processes.* London and New York: Psychology Press.
64 Massey, Ray (2009) 'There's no stopping "lycra lout" cyclists as prosecutions for running red lights plummet'. *Daily Mail.* 9 November.
65 CTC (2012) *Cycling and pedestrians briefing 4R.* On: www.ctc.org.uk
66 Dargay, J. and Hanly, M. (2007) 'Volatility of car ownership, commuting mode and time in the UK'. *Transportation Research Part A: Policy and Practice.* 41 (10), pp. 934-48.
67 DfT (2012) 'National travel survey'. Table NTS 0205. On: www.gov.uk.
68 Delbosc, A. and Currie, G. (2013) 'Causes of youth licensing decline: a synthesis of evidence'. *Transport Reviews.* 33 (3), pp. 271-90.
69 See for example: Goodwin, P. and Van Dender, K. (2013) '"Peak car" – themes and issues'. *Transport Reviews.* 33 (3), pp. 243-54, and:
IfMR (2013) *'Mobility Y: the emerging travel patterns of Generation Y'.* Munich: Institute for Mobility Research.

4 "Roads and airports benefit the economy"

70 Confederation of British Industry (2012) *Bold thinking: road report.* On: www.cbi.org.uk
71 The following parliamentary report provides a good summary of the evidence on the issues described over the next few paragraphs: SACTRA (1999) *Transport and the economy.* London: Standing Committee on Trunk Road Assessment.
72 For a good summary of the evidence see: Bhatta, S. D. and Drennan, M. P. (2003) 'The economic benefits of public investment in transportation: a review of recent literature'. *Journal of Planning Education and Research.* 22 (3), pp. 288-96.
73 Boarnet, M. G. (1996) 'The direct and indirect economic effects of transportation infrastructure'. Working Paper 340. University of California Transportation Center, Berkeley.
74 Fernald, J. G. (1999) 'Roads to prosperity? Assessing the link between public capital and productivity'. *American Economic Review.* 89 (3), pp. 619-38.
75 Cervero, R. (2009) 'Transport infrastructure and global competitiveness: balancing mobility and livability'. *The ANNALS of the American Academy of Political and Social Science.* 626 (1), pp. 210-25.
76 Sinnett, D., Williams, K., Chatterjee, K. and Cavill, N. (2011) 'Making the case for investment in the walking environment: a review of the evidence'. Living Streets, London. On: www.livingstreets.org.uk
77 Cairns, S., Newson, C. and Davis, A. (2010) 'Understanding successful workplace

travel initiatives in the UK'. *Transportation Research Part A: Policy and Practice.* 44 (7), pp. 473-94.
78 SACTRA (1994) *Trunk roads and the generation of traffic.* Report 11. London: Department for Transport Standing Advisory Committee on Trunk Roads Assessment.
79 Gorham, R. (2009) *Demystifying induced travel demand.* Report 1. Eschborn: Deutsche GesellschaftfurtTechnischeZusammenarbeit GmbH.
80 See for example: *Frontier Economics* (2011) 'Connecting for growth: the role of Britain's hub-airport in economic recovery. A report prepared for Heathrow'. London.
81 Smyth, M. and Pearce, B. (2007) *Aviation economic benefits: measuring the economic rate of return on investment in the aviation industry.* Economic report 8. Montreal: International Air Transport Association.
82 DfT (2013) *UK aviation forecasts.* Annex E1. London: Department for Transport.
83 *Frontier Economics* (2011) 'Connecting for growth: the role of Britain's hub-airport in economic recovery. A report prepared for Heathrow'. London.
84 DfT (2013) *Aviation policy framework.* London: Department for Transport.
85 Example: Smit, M., Koopman, M. and Faber, J. (2013) *The economics of airport expansion.* Report 13.7861.13. On: www.cedelft.eu
86 Lyons, G., Jain, J., Susilo, Y. and Atkins, S. (2012) 'Comparing rail passengers' travel time use in Great Britain between 2004 and 2010'. *Mobilities.* ISSN 1745-0101.
87 DfT (2011) 'Transport analysis guidance unit 3.4.1 The accidents sub-objective'. On: www.dft.gov.uk
88 For some shocking examples from the USA see: Ackerman, Frank and Heinzerling, Lisa (2004) *Priceless. On knowing the price of everything and the value of nothing.* New Press. New York.
89 See for example: Forster, Andrew (2013) 'We haven't cooked the books for HS2 appraisal, insists DfT's top mandarin'. *Local Transport Today.* 626, 12 July.
90 DfT (2009) *NATA refresh : Appraisal for a sustainable transport system.* London: DfT
91 DfT (2013) Project Database. Project: 'WebTAG: international benchmarking'. On: www.gov.uk
92 Wenban-Smith, Alan (2013) 'If our transport appraisal methods are so smart, why is our transport system so poor?' *Local Transport Today.* 19 April.

5 "All we need is better public transport"

93 Gardiner, Clare (2014) 'Stagecoach warning over new rules for bus firms'. *The Scotsman.* 15 August.
94 Association of Train Operating Companies (2013) 'Growth and prosperity: How franchising helped transform the railway into a British success story'. On: www.atoc.org
95 HM Government (2010) *The Coalition: Our programme for government.* 20 May. p 31.
96 Transport for London (2012) *Travel in London Report 5.*
97 DfT (2012) 'National travel survey'. Table TSGB 0101.
98 Knowles, R. D. (1996) 'Transport impacts of greater manchester's metrolink light rail system'. *Journal of Transport Geography.* 4 (1), 1-14.
99 Fastrack Delivery Executive (2006) 'Fastrack - the first six months'. On: www.go-fastrack.co.uk
100 DfT (2012) 'Economic case for high speed rail; updated appraisal of transport user benefits and wider economic benefits'. London. Department for Transport
101 Parkhurst, G. (2000) 'Influence of bus-based park and ride facilities on users' car traffic'. *Transport Policy.* 7 (2), 159-72.
102 Minnerva Ltd & MVA Consultancy (2013) 'Scotland-wide older and disabled persons concessionary bus scheme - further reimbursement research'. Transport for Scotland. www.transportscotland.gov.uk
103 DfT (2012) 'National travel survey'. Table Bus 0103
104 Eurostat Table tsdtr210 (Modal split of passenger transport) shows that Britain (including the big increase in London) suffered the third worst decline in bus use in Western Europe since 1991. A resurgence in cycling was a major factor in the two countries which declined more, Germany and the Netherlands.
105 DfT (2012) 'Transport statistics'. Table BUS0405.
106 Rye, T. (2008) 'Mind the Gap! The UK's record in European perspective'. In: Docherty, I. and Shaw, J. eds (2008) *Traffic Jam: Ten Years of 'Sustainable' Transport in the UK.* Bristol: Policy, p. 220 and footnote 2.
107 Stagecoach (2013) Tyne and Wear Integrated Transport Authority. 'Proposal for a quality contracts scheme in Tyne and Wear: consultation response of Stagecoach Group Plc'. On: www.stagecoachbus.com

108 Competition Commission (2011) *Local bus services market investigation*. 'Provisional findings report'. On: www.competition-commission.org.uk
109 See for example: Forster, Andrew (2011) 'Bus industry big guns take aim at Competition Commission findings'. *Local Transport Today*. 576. 29 July.
110 Competition Commission (2011) *Local bus services market investigation*. 'A report on the supply of local bus services in the UK (excluding Northern Ireland and London)'. On: www.competition-commission.org.uk
111 Wolmar, Christian (1999) *Stagecoach: A classic rags to riches tale from the frontiers of capitalism*. Orion.
112 Monopolies and Mergers Commission (1995) *The supply of bus services in the north-east of England*. Summary On: http://archive.today/fUrZb
113 Wolmar, C. (2005) *On the wrong line: How ideology and incompetence wrecked Britain's railways*. Rev. and updated ed. London: Aurum.
114 Butcher, L. (2013) *Rail Passenger Franchises*. Report: 1343. House of Commons Library. On: www.parliament.uk:
115 Webster, Ben (2004) 'Most punctual commuter trains are the services run by the State'. *The Times*. 18 November.
116 Topham, Gwynn (2013) 'East Coast rail service costs taxpayers less than private lines, report reveals'. *The Guardian*. 18 April.
117 Bowman, A., Folkman, P., Froud, J., Johal, S., Law, J., Leaver, A., Moran, M. and Williams, K. (2013) *The Great Train Robbery: The economic and political consequences of rail privatisation*. Centre for Research on Socio-Cultural Change. On: www.cresc.ac.uk
118 DfT (2013) 'Rail franchising future programme'. Press release. 31 January. On: www.gov.uk
119 Association of Train Operating Companies (2013) *Growth and prosperity: How franchising helped transform the railway into a British success story*. On: www.atoc.org
120 Bowman, A., Folkman, P., Froud, J., Johal, S., Law, J., Leaver, A., Moran, M. and Williams, K. (2013) *The Great Train Robbery: The economic and political consequences of rail privatisation*. Centre for Research on Socio-Cultural Change. On: www.cresc.ac.uk
121 Total debt for the financial year shown as £32.7bn. 'Network Rail Infrastructure Finance PLC Annual Report and Accounts Year ended 31 March 2013'. On: www.networkrail.co.uk.
122 Based on 'National travel survey' (2011) Table TSGB 0104 (multiplied by the population of Great Britain). The car trips projected to be saved by HS2 come from: DfT (2012). 'Economic case for high-speed rail: Updated appraisal of transport user benefits and wider economic benefits'. London: Department for Transport. The overall cost of HS2 is based on: HM Treasury (2013) *Investing in Britain's Future*. London. On: www.gov.uk

6 "Car ownership isn't a problem – only car use"

123 See for example: Van Acker, V. and Witlox, F. 2010. 'Car ownership as a mediating variable in car travel behaviour research using a structural equation modelling approach to identify its dual relationship'. *Journal of Transport Geography*, 18(1), pp. 65-74
124 DfT (2011) Tables TRA 0101 and VEH 0153
125 Commission for Integrated Transport (2000). *European best practice in transport – the German example*. Available on the National Archives website.
126 The car ownership comparisons were for 2009, the trips for 2010 and the distance for 2011. Sources: Car ownership: Eurostat (2009). Modal shares: DfT 'National travel survey' (Tables 0204 and 0101). Distance: DeStatistik.
127 DfT (2011) 'National travel survey'. Table NTS0702.
128 DfT (2011) Transport statistics Table NTS0205.
129 DfT (2006) 'Attitudes to car use: London.' Available on the National Archives website.
130 Scottish Executive (1999) *Why people don't drive cars*. Scottish Executive, Central Research Unit.
131 Melia, S. (2009) 'Potential for Carfree Development in the UK'. PhD, University of the West of England. On: www.stevemelia.co.uk.
132 A special edition of the journal *Transport Reviews* was devoted to this subject in 2009: Volume 29, Issue 3.
133 LTT (2009) 'Kent tackles "discredited ideologies" of residential parking restrictions'. *Local Transport Today*. 16 January (511).
134 DETR (2001) 'Planning policy guidance note 13: Transport'. The Stationery Office.
135 CLG (2011) *Pickles and Hammond to end the war on the motorist*. On: www.gov.uk

136 Ross, L., Greene, D. and House, P. (1977) 'The "false consensus effect": An egocentric bias in social perception and attribution processes'. *Journal of Experimental Social Psychology*. 13 (3), pp. 279-301.

137 WSP (2012) 'Does Car Ownership Increase Car Use?' The Berkeley Group. On: www.berkeleygroup.co.uk.

139 Islington Borough Council (2012) 'Development management policies, Topic paper: transport'. On: www.islington.gov.uk

140 Compare, the carefully nuanced conclusions of Le Vine and Jones (2012) in *On the Move*, RAC Foundation, with the more political overview written by the sponsoring organisations.

141 RAC Foundation (2005) *Motoring towards 2050: Parking in transport policy*.

142 DfT (2006) *Attitudes to car use*. London. Available on the National Archives website.

143 DfT (2006) *Attitudes to car use*. London. Chapter 4. Available on the National Archives website

144 Melia, S. (2014) 'Carfree and low car development'. In: Ison, S.G. and Mulley, C. eds (2014) *Parking: issues and policies*. Emerald Insight, Bingley, West Yorkshire.

145 Guo, Z. (2013) 'Does residential parking supply affect household car ownership? The case of New York City'. *Journal of Transport Geography*. 26 (0), pp. 18-28.

146 Thomas, R (2012) 'London bucks trend with jump in car-free household numbers'. *Local Transport Today*. Issue 612

147 Melia, S. (2011) 'Students car use and its effect on environmental attitudes'. Bristol: University of the West of England. On: eprints.uwe.ac.uk/14424/

148 Hass-Klau, C., Crampton, G.R. and Ferlic, A. (2007) 'The effect of public transport investment on car ownership'. Environmental and planning, UK; University of Wupperthal.

149 Lucas, K. and Jones, P. (2009) *The car in British society*. London: RAC Foundation. Figure 2.2

150 National Centre for Social Research (2011) *British social attitudes survey*. Report: 28. London: Sage.

7 "You'll never get people over here cycling like the dutch"

151 DfT (2007) *Manual for streets*. Thomas Telford Publishing: London.

152 Gallup, Hungary (2011) 'Future of transport: analytical report'. Report: 312. On: ec.europa.eu

153 Shown as 9 per cent on a different measure in: European Commission (2011) 'The promotion of cycling'. Brussels: Directorate General for Internal Policies.

154 Continental statistics from: European Platform on Mobility Management, Modal Split Tool. On: epomm.eu/tems/
London data from: Transport for London (2011) 'Travel in London supplementary report: London travel demand survey'.

155 2011 census – travel to work only, excluding home workers.

156 Comparisons made on: www.worldweatheronline.com

157 See for example: Pooley, C., Tight, M., Jones, T., Horton, D., Scheldeman, G., Jopson, A., Mullen, C., Chisholm, A., Strano, E. and Constantine, S. (2011) 'Understanding walking and cycling: summary of key findings and recommendations'. Lancaster University.

158 Melia, S. (2009) 'Potential for carfree development in the UK'. Unpublished PhD, University of the West of England. p. 228. On: www.stevemelia.co.uk.

159 These articles are all on: www.stevemelia.co.uk/articles.html.

160 Melia, S. (2008) 'Neighbourhoods should be made permeable for walking and cycling but not for cars'. *Local Transport Today*. 23 January.

161 See for example: Wardman, M., Tight, M. and Page, M. (2007) 'Factors influencing the propensity to cycle to work'. *Transportation Research Part A: Policy and Practice*. 41 (4), pp. 339-50.

161i Cycling Embassy of Denmark (2010) *Bicycle statistics of Denmark*. On: www.cycling-embassy.dk
Mobilität in Deutschland (2012) 'National transportation surveys activities in Germany'. On: www.mobilitaet-in-deutschland.de

161ii Source: Netherlands Ministry of Infrastructure and the Environment (direct communication)

161iii See for example: Pucher, J., & Buehler, R. (2008). 'Making cycling irresistible: Lessons from the Netherlands, Denmark and Germany'. *Transport Reviews*. 28(4), 495-528.

161iv Cruz, Michel (2011) 'Revolution – the story of cycling success in Seville'. *Cycling Mobility* 1/2011, 72-9

161v José I. Castillo-Manzano & Antonio Sánchez-Braza (2013) 'Can anyone hate the bicycle? The hunt for an optimal local transportation policy to encourage bicycle usage'. *Environmental Politics*. 22:6, 1010-28,

161vi Tapp, Alan (2013) 'Trends in British attitudes to Cycling: A tracker survey of the GB population'. Unpublished conference presentation. On: www2.uwe.ac.uk

161vii Yougov (2013) *Profile of voters February 2013*. On: cdn.yougov.com

161viii As Easy as Riding a Bike (2013) 'No surrender' – the damaging, enduring legacy of the 1930s in British cycle campaigning'. 15 February. On: aseasyasridingabike.wordpress.com

161ix www.margaretthatcher.org

162 Grous, A. (2011) 'The British cycling econmy: gross cycling product report' On: eprints.lse.ac.uk/38063

163 Bristol Post (2014) 'Conservative raises concerns over cash commitment to cycling!' 16 July.
Following this story, I went on Radio Bristol to explain that the proposed budget of £35m over 5 years would equate to around 5 per cent of transport spending per person nationally. In a city where cycling accounts for 8 per cent of journeys, this was hardly excessive.

164 This phrase became something of a catchphrase to summarise the attitudes of that government towards transport policy. The original reference: Thatcher, Margaret (1990) 'Speech presenting 1989 Better environment awards for industry'. Royal Society of Arts, Central London. 16 March. On: www.margaretthatcher.org

165 There is a lively debate conducted on these issues (marred by some unfortunate personal abuse) on several cycling blogs, including aseasyasridingabike.wordpress.com and departmentfortransport.wordpress.com reflecting some differing views about who said and did what in the past.

166 CIHT, Bicycle Association, CTC and DfT (1996) 'Cycle friendly infrastructure guidelines for planning and designing'.

167 DfT (2012) 'Local transport note 1/12: Shared use routes for pedestrians and cyclists'. On: www.gov.uk

168 DfT (2007) *Manual for Streets*. Thomas Telford Publishing: London.

169 DfT (2008) 'Local transport note 2/08: Cycle design infrastructure guide'.

170 Pooley, C., Tight, M., Jones, T., Horton, D., Scheldeman, G., Jopson, A., Mullen, C., Chisholm, A., Strano, E. and Constantine, S. (2011) 'Understanding walking and cycling: Summary of key findings and recommendations'. Lancaster University.

171 DfT (2013) Statistical release – 'National travel survey 2012'. On: www.gov.uk

172 Hembrow, D. and Wagenbuur, M. (2012) 'Campaign for sustainable safety, not strict liability'. On: www.aviewfromthecyclepath.com

173 ONS (2013) Census Table CT0015. Home workers and non-workers excluded.

174 DfT (2013) 'Local area walking and cycling'. London.

175 www.cycling-embassy.org.uk

176 CTC (2012) 'Cycle-friendly design and planning'. On: www.ctc.org.uk

177 Mayor of London (2013) 'The mayor's vision for cycling – an Olympic legacy for all Londoners'. Greater London Authority.

178 See for example: BBC News (2014) 'Bristol to get first "Dutch-style" segregated cycle lanes'. 14 February. On: www.bbc.co.uk

179 Pucher, J. and Buehler, R. (2008) 'Making cycling irresistible: Lessons from the Netherlands, Denmark and Germany'. *Transport Reviews*. 28 (4), pp. 495-528.

180 The path is a two-way path. Another excuse sometimes given for the lack of priority is that drivers will not be looking out for cyclists moving in both directions. This is a short-term problem of driver familiarity: once many such paths have been installed, drivers will become familiar with them. In the meantime, the Dutch have another solution to the short-term problem – the raised table similar to the one shown in Figure 12.21 (page 139).

8 "The car can be a guest in our streets"

181 CIHT & DfT (2010) *Manual for streets 2 – Wider application of the principles*. Chartered Institute of Highways and Transportation; Department for Transport.

182 DfT (2007) *Manual for Streets*. Thomas Telford Publishing: London.

183 DfT (2011). 'Local transport note 1/11 – shared space'. Department for Transport.

184 Moody, S. and Melia, S. (2013) 'Shared space – implications of recent research for transport policy'. Proceedings of the Institution of Civil Engineers. On: eprints.uwe.ac.uk/17937/

185 Gerlach, J. Methorst, R., Boenke, D. and Levens, J. (2008) 'Sense and Nonsense in Shared Space'. *Feitsberaad* On: www.fietsberaad.nl

186 Hamilton-Baillie, B. (2008) 'Towards shared space'. *Urban Design International*. 13 (2), pp. 130-8.

187 See for example: Cassini, M. (2006) 'In your car no one can hear you scream! Are traffic controls in cities a necessary evil?' *Economic Affairs*. 26 (4), pp. 75-8.
188 Hamilton-Baillie, B. (2008) 'Should pedestrian guard rails be removed in urban areas to make streets more "dangerous" and pedestrians and drivers more careful, thereby reducing accidents?' *Transportation Professional*. Jan/Feb p. 4.
189 Hamilton-Baillie, B. (2008) 'Towards shared space'. *Urban Design International*. 13 (2), pp.130-8. This paper claims that several British local authorities have implemented shared space schemes as a "key policy component" to achieve "modal shift to walking and cycling".
190 Kent County Council (2012) 'Ashford ring road: Report on three years of two-way operation'.
191 Coventry City Council (2012) 'Public Report 5: Response to petitions received concerning "shared space concept"'. 28 February.
192 MVA Consultancy (2009) 'DfT shared space project stage 1: Appraisal of shared space'. Available on the national archives website.
193 DfT (2011) 'Local transport note 1/11 – shared space'. Department for Transport; TSO.
194 MVA Consultancy (2010) 'Designing the future – shared space: Operational research. Report for the Department for Transport'.
195 DfT (2011) 'Local transport note 1/11 – shared space'. Department for Transport; TSO. Section 2.11
196 MVA Consultancy (2010) 'Designing the future– shared space: Qualitative research'. Available on the national archives website.
197 Guide Dogs for the Blind. 'Shared surface streets perceived as 'no go' areas by blind and partially sighted people says new national survey'. On: www.guidedogs.org.uk
198 A short introduction by the filmmaker, Sarah Gayton, is available on *YouTube*.
199 The term 'Belisha beacon' was named after the transport minister Leslie Hore-Belisha, who approved their use in the 1930s.
200 See for example: Trayner, David (2013) '"Confusing" crossings cost £241,000'. *Southend Echo*. 28 June. On: www.echo-news.co.uk and: Forster, Andrew (2014) 'Blind groups press council to scrap courtesy crossing plans'. *Local Transport Today*. Issue 656. 14 April. This article refers to Solihull in the West Midlands.
201 See for example: MVA Consultancy (2009) 'DfT shared space project stage 1: Appraisal of shared space'. p. 2. Available on the national archives website.
202 See for example: Sustrans (2006) 'Shoppers and how they Travel'. Bristol.
203 CIHT & DfT (2010) *Manual for streets 2 – wider application of the principles*. London: Chartered Institute of Highways and Transportation; Department for Transport.
204 See for example: Colin Buchanan and Partners (2004) 'Report for Aberdeen City Council Union Street Project'. And: Whitehead, T., Simmonds, D. and Preston, J. (2006) 'The effect of urban quality improvements on economic activity'. *Journal of Environmental Management*. 80 (1), pp. 1-12.
205 Building Design Partnership (2006) 'Nelson Town Centre Masterplan'. Pendle Borough Council.
206 LGN (2012) LGN Street Design Awards. Winner: Pendle Borough Council, Nelson Town Centre. On: awards.lgn.co.uk
207 See for example: Pooley, C., Tight, M., Jones, T., Horton, D., Scheldeman, G., Jopson, A., Mullen, C., Chisholm, A., Strano, E. and Constantine, S. (2011) 'Understanding walking and cycling: Summary of key findings and recommendations'. Lancaster University.
208 Litman, T. (2012) 'Land use impacts on transport'. Victoria Transport Policy Institute. On: www.vtpi.org
209 Pooley, C., Tight, M., Jones, T., Horton, D., Scheldeman, G., Jopson, A., Mullen, C., Chisholm, A., Strano, E. and Constantine, S. (2011) 'Understanding walking and cycling: Summary of key findings and recommendations'. Lancaster University.
210 Frank, L. D., & Hawkins, D. (2008) 'Giving pedestrians an edge – Using street layout to influence transportation choice'. Canada Mortgage and Housing Corporation: Ottawa.
211 Rietveld, P. and Daniel, V. (2004) 'Determinants of bicycle use: do municipal policies matter?' *Transportation Research Part A: Policy and Practice*. 38 (7), pp. 531-50. And: Schepers, P., Heinen, E., Methorst, R. and Wegman, F. (2013) 'Road safety and bicycle usage impacts of unbundling vehicular and cycle traffic in Dutch urban networks'. *European Journal of Transport and Infrastructure Research*. 13 (3), pp. 221-38.
212 Congress for the New Urbanism (2001) *Charter for the new urbanism*. On: www.cnu.org/charter

213 See for example: DfT (2005) 'Home zones: Challenging the future of our streets'. Department for Transport: London.

9 "We are building too many flats"

214 Echenique, Marcial (2010) 'Does high density development make travel more sustainable?' *CABE Sustainable Places*. Available on National Archives website.

215 Town and Country Planning Association (2014) *Garden City Principles*. On: www.tcpa.org.uk

216 Morton, A. (2011) 'We must trust people's instincts'. *Planning Magazine*. 5 December.

217 ONS (2014) Electoral Wards/Divisions. On: www.ons.gov.uk

218 Pateman, T. (2010) 'Rural and urban areas: Comparing lives using rural/urban classifications'. Report 43 Regional Trends. London: Office for National Statistics.

219 Statistics from the 2011 census for wards in England and Wales.

220 An entire edition of the academic journal *Transport Reviews* was dedicated to these questions in 2009. See for example: Næss, P. (2009) 'Residential self-selection and appropriate control variables in land use: travel studies'. *Transport Reviews*. 29 (3), pp. 293-324.

221 SOLUTIONS (Sustainability Of Land Use and Transport In Outer Neighbourhoods) On: www.suburbansolutions.ac.uk

222 Lambert, C., Griffiths, R., Oatley, N. and Smith, I. (1999) 'On the edge: The development of Bristol's north fringe'. Working Paper 9. University of the West of England.

223 Melia S. (2005) 'Regeneration & Renewal'. 13 May. On: www.stevemelia.co.uk

224 SOLUTIONS project. Graph derived from the raw data. On: www.suburbansolutions.ac.uk.

225 Horswell, M. and Barton, H. (2010) 'Active travel patterns and neighbourhood accessibility. Summary findings from SOLUTIONS WP12'. Note this graph was drawn directly from the raw data and differs slightly from the one shown in that paper. On: www.suburbansolutions.ac.uk.

226 Melia, S., Barton, H. and Parkhurst, G. (2011) 'The paradox of intensification'. *Transport Policy*. 18 (1), 46-52.

227 Density statistics from: CLG (2011) 'Land use change statistics in England: 2010 (provisional estimates)'. On: www.gov.uk Distance driven (Great Britain) from 2011 'National travel survey' (Table 101).

228 Evans, A. and Unsworth, R. (2012) 'Housing densities and consumer choice'. *Urban Studies*. 49(6), pp. 1163-1177.

229 Transport Scotland (2012) Statistical Bulletin – 'National travel survey' 2011/2012: Table 5. On: www.transportscotland.gov.uk

230 Melia, S. (in press) 'Evaluating policy impact – The built environment and travel behaviour'. In: Spotswood, Fiona eds *Beyond Behaviour Change*. Policy Press.

231 CLG (2010) 'Household projections, 2008 to 2033, England'. On: www.gov.uk

232 Hope, Christopher (2011) 'Planning reforms: David Cameron aide privately told builders new NPPF would "trigger more development"'. *Daily Telegraph*. 23 September. On: www.telegraph.co.uk

233 Communities & local government (2010) Press Release. 9 June.

234 Melia, S. (2010) 'Urban intensification and problems – real and imagined'. *Town and Country Planning*. July/August pp. 341-45.

235 CLG (2014) 'Housebuilding: permanent dwellings completed, by house and flat, number of bedroom and tenure, England'. Table 254 On: www.gov.uk

236 CLG (2013) 'Land use change: Proportion of new dwellings on previously developed land, England, 1989 to 2011'. Table P211. On: www.gov.uk.

237 Tables QS112EW and QS406EW. On: www.nomisweb.co.uk

238 Date except England and Scotland came from: Eurostat (2013) 'Distribution of population by degree of urbanisation, dwelling type and income group'. On: epp.eurostat.ec.europa.eu
Data for England from: CLG (2011) 'English housing survey home report 2010'. Annex AT1.5. Available on www.gov.uk.
Date for Scotland from: National Records of Scotland (2012) 'Estimates of households and dwellings in Scotland, 2011'. On: www.gro-scotland.gov.uk

239 CLG's household projections do not show household size in sufficient detail, but they do show average household size continuing to decline to 2021.

240 2011 census table QS402EW – refers to occupied properties only.

241 See for example: Sefton Metropolitan Borough Council (2013) 'Housing technical paper August 2013'. on www.sefton.gov.uk And: Harlow Borough Council (2009) 'Harlow's local development framework core

strategy workshop October 2009'.
On: www.harlow.gov.uk
242 Graeme Browne, director of Shelter Scotland, quoted in: McVeigh, Tracey (2009) '30 years on, the right to buy revolution that still divides Britain's housing estates'. *The Observer*. 6 December.
243 Reeve, Kesia (2011) 'The hidden truth about homelessness: experiences of single homelessness in England'. *Crisis*. On: www.crisis.org.uk
244 Capital letters are usually used where describing the New Towns built under the New Towns Act.
245 Savills (2014) 'Market survey: UK agricultural land 2014'. On: www.savills.co.uk
246 GVA (2013) 'Research report: development outlook summer 2013'. On: www.gva.co.uk
247 'National travel survey.' (2012) Table NTS9903. On: www.gov.uk
248 CLG (2009) 'Planning policy statement: eco-towns, a supplement to planning policy statement 1'. London: Communities and Local Government.
249 ONS (2014) census 2011. Tables QS701EW and QS702EW – home workers and non-workers excluded. Electoral wards or Output areas were used in this table. Some of them may not exactly correspond to the boundaries of each settlement.
250 DfT (2008) 'Building sustainable transport into new developments: a menu of options for growth points and eco-towns'. On: www.tcpa.org.uk
251 TCPA and CLG (2008) 'Eco-towns communities worksheet'. London: Town and Country Planning Association. On: www.tcpa.org.uk
252 Meghji, Shafik (2008) 'Eco-towns must not promote car use, say campaigners'. 9 May. On: www.planningresource.co.uk
253 CLG (2014) 'Locally led garden cities'. On: www.gov.uk
254 Town and Country Planning Association (2011) 'Re-imagining garden cities for the 21st century'. On: www.tcpa.org.uk
255 See for example: Hallam Land Management, Persimmon, Redrow and Taylor Wimpey (2010) 'East Devon new community phasing strategy'. On: www.eastdevon.gov.uk. And: Northstowe Joint Development Control Committee (2012) 'Northstowe development framework document'. On: www.scambs.gov.uk.
256 South West Regional Assembly (2007) 'Panel report draft regional spatial strategy for the southwest'. On: www.southwest-ra.gov.uk. This report recommended a further extension to Cranbrook new settlement before building had even begun. The Regional Spatial Strategies were abolished by the Coalition government in 2010, but the inspector considering the East Devon Local Plan in 2013/14 also recommended the district council to search again for new sites for more housing: the proposal to extend Cranbrook is very likely to re-emerge before too long.
256i Labour Party (2014) 'The Lyons housing review: Mobilising across the nation to build the homes our children need'. On: www.yourbritain.org.uk
257 South Hams District Council (2005) 'Core strategy – preferred options stage'. On: www.southhams.gov.uk
258 South Hams Against Rural Destruction (2006) 'Evidence to the examination in public into the South Hams core strategy'.
259 Hickman, R. and Banister, D. (2008) 'Transport and reduced energy consumption: The role of urban planning' in 40th Universities Transport Study Group Conference. January.
And: Headicar, P. and Curtis, C. (1994) 'Residential development and car-based travel: does location make a difference'. *Seminar C: Environment Issues*. September. PTRC European Transport Forum.
260 DfT (2014) National travel statistics. Tables NTS 0401. On: www.gov.uk
261 2011 census tables CS0010001 and CS0300010. On: www.nomisweb.co.uk
262 Piggott, G (2007) 'Commuting in London, DMAG briefing 2007-03. Greater London authority data management and analysis group'. On: legacy.london.gov.uk
263 Halcrow Fox (2000) 'Spatial strategy for the south west'. Southwest Regional Planning Conference. Incorporated as Background Paper 3 to the Regional Planning Guidance. On: www.southwest-ra.gov.uk
264 Jain, J. and Lyons, G. (2008) 'The gift of travel time'. *Journal of Transport Geography*. 16 (2), pp. 81-9.
265 Hickman, R. and Banister, D. (2007) 'Transport and reduced energy consumption: the role of urban planning'. Working Paper 1026. Oxford University Transport Studies Unit.
266 Ewing, R. and Cervero, R. (2010) 'Travel and the built environment: A meta-analysis'. *Journal of the American Planning Association*. 76 (3), pp. 265-94.

10 Summary: myths, values and challenges

267 See for example the government's definition of 'sustainable development' within (2012) 'National planning policy framework'. On: www.gov.uk.
268 Goodall, Chris (2013) *Sustainability – all that matters*. Abingdon: Hodder Education.

11 Four options for traffic in towns

269 Buchanan, C., Douglas (1964) *Traffic in towns: The specially shortened edition of the Buchanan report*. Harmondsworth: Penguin Books.
270 DfT (2012) 'Transport statistics Great Britain', Table VEH0153, divided by mid-year population statistics from: ONS (2011) 'Mid-1971 to Mid-2010 Population Estimates: United Kingdom; estimated resident population for constituent countries and regions'. Both on: www.gov.uk
I am grateful to Professor Phil Goodwin who brought this to my attention, in one of his fascinating presentations about 'peak car'.
271 Openstreetmap data is crowd sourced, ie it is created by thousands of volunteers. It is made available under the Open Database License. See www.openstreetmap.org for more information on this valuable resource.
272 DfT (2014) Transport statistics Table CGN0206a
273 Bristol City Council (2012) 'Public realm and movement framework consultation draft'. On: www.bristol.gov.uk
274 Department for Business, Innovation and Skills and DfT (2014) 'UK government fast tracks driverless cars'. Press release. 20 July. On: www.gov.uk
275 Brundtland, G.H. (1987) *Our common future: World commission on environment and development*. Oxford: Oxford University Press.
276 DEFRA (1999) 'A better quality of life – strategy for sustainable development for the United Kingdom, London'. Available on the National Archives website.
277 Eurostat (2011) 'Modal split of passenger transport'. On: epp.eurostat.ec.europa.eu
278 See for example: Highways Agency (2013) 'Post opening project evaluation (POPE) of major schemes meta-analysis 2013. Executive summary'. On: www.highways.gov.uk

12 European cities: inspiration and similarities

279 The section which follows draws on a mixture of my own notes, made on my visit to Freiburg in 2006, numerous sources provided by Freiburg city council, then and more recently, and on the following three sources, which also draw heavily on data from the city council:
Beim, M. and Haag, M. (2010) 'Freiburg's way to sustainability: the role of integrated urban and transport planning'. In Schrenk, M., Popovich, V. and Zeile, P. eds (2010) *Proceedings of real corp*. Vienna, May.
Buehler, R. and Pucher, J. (2011) 'Sustainable transport in Freiburg: lessons from Germany's environmental capital'. *International Journal of Sustainable Transportation*. 5 (1), pp. 43-70.
FitzRoy, F. and Smith, I. (1998) 'Public transport demand in Freiburg: why did patronage double in a decade?' *Transport Policy*. 5 (3), pp. 163-173.
The statistics quoted all come from Freiburg City Council and/or VAG, the public transport operator.
280 This was the conclusion reached by FitzRoy, F. and Smith, I. (1998) 'Public transport demand in Freiburg: why did patronage double in a decade?' *Transport Policy*. 5 (3), pp. 163-173.
281 Fiets Beraad (2010) 'Bicycle policies of the European principals: continuous and integral'. Report: 7. Rotterdam.
282 City of Freiburg and Academy of Urbanism (2010) Freiburg charter. On: www.wulf-daseking.de
283 The trip numbers, which don't include walking trips, were taken from Heller (2002) divided by population statistics from Freiburg.de.
Heller, P.W. (2002) 'Sustainable human development in a medium-sized city: The example of Freiburg, Germany'. In Sassen, S. ed (2002). *Encyclopedia of Life Support Systems*. Oxford.
284 Dale, Steven (2011) 'Groningen, bicycle capital of the world'. On: gondolaproject.com
285 Fiets Beraad (2010) 'Bicycle policies of the European principals: continuous and integral'. Report: 7. Rotterdam.
286 What follows draws on my own notes, including an interview with the then senior transport planner, Cor Van der Klaauw, information provided by Groningen City

Council, Fiets Beraad[285] and, for the story of the city centre traffic plan: Tsubohara, S., (2007) 'The effect and modification of the traffic circulation plan (VCP) – Traffic planning in Groningen in the 1980s'. Report: 317. Groningen: Urban and Regional Studies Institute.
All the city statistics come directly or indirectly from Groningen City Council. National statistics are from statline.cbs.nl

287 Groningen city council retail survey 2010/11
288 Email to and from Frank Broersma, Groningen Chamber of Commerce. 20 December 2011.
289 groningen.nl (2013) 'Hoe duurzaam is de Groningse student?' 19 March.
290 volkskrant.nl (2012) 'Groningen is 40 miljoen kwijt door Regiotram'. 17 October.
291 Most of what follows comes from three study visits from 2010 to 2012 and material provided by Grand Lyon, Sytral (the public transport operator) and Région Rhone-Alps, with some reference to the following:
Bouf, D. and Hensher, D.A. (2007) 'The dark side of making transport irresistible: The Example of France'. *Transport Policy*. 14 (6), pp.523-532.
D'Arcier, F. (2003) 'Urban transport in France: Moving to a sustainable policy'. *Senshu Diagaku Shakai Kagaku Kenyo Geppo*. 481 pp. 11-27.
Harman, R., L'Hostis, A. and Menerault, P. (2007) 'Public transport in cities and regions: facing an uncertain future?' In Booth, P., Breuillard, M., Fraser, C. and Paris, D. eds (2007) *Systems of Britain and France: A Comparative Analysis* pp. 188-204. London: Routeledge.
292 From: www.sytral.fr
293 Région Rhône-Alpes (2007) On: www.rhonealpes.fr
294 This section is taken from: Calibre, E., Simon, E. and Turpin, M. (2007) 'Berges du Rhône'. Lyon: Université Lumière Lyon.
295 See: www.dailymotion.com/video/x21n0o_berges-du-rhone_news
296 C4ndide (2012) 'Les Berges du Rhône, l'envers du décor'. On: lyon.citycrunch.fr
297 D'Huissel, Sylvain (2013) 'Étalement urbain: le SNAL veut lutter contre les idées recues'. 19 March. On: www.lyonpoleimmo.com
In this article, the regional developers' association presents a report claiming that new detached houses are not contributing (very much) to urban sprawl around Lyon, and attempts to restrain them are hitting the middle classes with higher house prices.
298 Axel, Gilbert (2011) 'Unités urbaines 2010 : accélération de l'étalement urbain'. *Insee Rhône-Alpes - La Lettre Résultats*. 147. June. On: www.insee.fr/rhone-alpes
299 483 per 1000 people in 2006 according to: epp.eurostat.ec.europa.eu
300 Sytral (2007) 'Enquête ménages déplacements'. On: www.sytral.fr
301 Observatoire des Territoires (2010) 'L'évolution des parts modales des déplacements domicile-travail'. On : www.observatoire-des-territoires.gouv.fr
302 van Wissen, Leo (2012) 'Demografische veranderingen op het platteland'. Presentation to symposium het platteland als consumptieruimte. 12 April. *Rijksuniversitiet Groningen*. On: www.knag.nl
This presentation showed falls in most other age groups, particularly between 30 and 45.
303 Sytral (2008) 'Enquête ménages déplacements - principaux résultats pour l'agglomération Lyonnaise'. On: www.urbalyon.org
304 Stadt Freiburg (2013). Searchable tables from 2003 onwards are available from: fritz.freiburg.de
305 'Sustainable traffic policy of the city of Freiburg'. Unpublished presentation sent to the author by City of Freiburg Transport Department.

13 Carfree developments

306 Some quite shocking TV footage of the events is available online. A search on 'Walen Buiten Louvain' will produce several examples.
307 This quote and much of what follows draws on: Remy, Jean. (2007) *Louvain-la-Neuve, une manière de concevoir la ville: Genèse et évolution*. New edition. Louvain-la-Neuve: Presses universitaires de Louvain. On: books.openedition.org.
Other sources include my own notes made in 2007 and: Frankignoulle, Pierre and Bodson, Edith (1996) 'Le campus universitaire comme espace public: des représentations aux pratiques'. *Études de communication*.
On: edc.revues.org/2453
308 Reproduced courtesy of l'Université Catholique de Louvain
309 UCL (2013) 'L'UCL à Louvain-la-Neuve: Modernité et tradition d'une ville piétonne'. On: www.uclouvain.be

310 See : Jacobs, Jane (1961) *The Death and Life of Great American Cities*. New York: Random House.
311 See: Mcallister, Gilbert (1946) 'Planning New Towns'. *The Spectator*. 5 April. p. 8 On: archive.spectator.co.uk
312 Colleyn Mathieu (2013) 'Bruxelles veut le RER "maintenant, tout de suite"' *La Libre Belgique*. 3 October.
313 Hermia, J-P. and Eggerickx, T (2007) 'Louvain-la-Neuve : une démographie atypique'. *La Gazette du Sped*. 21, p. 1-2
314 Hermia, J-P. and Eggerickx, T (2007) 'Louvain-la-Neuve : une démographie atypique'. *La Gazette du Sped*. 21, p. 1-2
315 Rémy, Jean. (2007) *Louvain-la-Neuve, une manière de concevoir la ville: Genèse et évolution*. New edition. Louvain-la-Neuve: Presses universitaires de Louvain.
316 Nützel, M. (1993) 'Nutzung und Bewertung des Wohnumfeldes in Großwohngebieten am Beispiel der Nachbarschaften U und P'. *Nürnberg-Langwasser*. 119. UniversitatBayreuth.
317 UNICEF (2013) 'Child well-being in rich countries. A comparative overview'. Innocenti Report Card 11. Florence: Innocenti Research Centre. On: www.unicef.org.uk
318 Field, Simon (2011) 'Vauban case study, Freiburg, Germany'. Institute for Transportation and Development Policy. On: www.itdp.org
319 Freiburg city council cited in: Moore, Tristana (2009) 'Heroes of the environment: residents of Vauban'. On: content.time.com
320 Nobis, C. (2003) 'The impact of car-free housing districts on mobility behaviour – case study'. In Beriatos, E., Brebbia, C. A., Coccossis, H. and Kungolos, A. eds (2003) International Conference on Sustainable Planning and Development. Skiathos, Greece: WIT.
321 Kuntz, A. (2006) Geschichten vom Vauban. Freiburg: Self-published.
322 Ornetzeder, M., Hertwich, E.G., Hubacek, K., Korytarova, K. and Haas, W. (2008) 'The environmental effect of car-free housing: A case in Vienna'. *Ecological Economics*. 65 (3), pp. 516-30.
323 Personal correspondence (2005) Fergus Allen, Dunedin Canmore. 9 December.
324 See: Eastwood, M. (2008) 'Slateford Green transport study'. Edinburgh: Dunedin Canmore Housing Association.
325 See for example: Hass-Klau, C. (1993) 'Impact of pedestrianization and traffic calming on retailing: A review of the evidence from Germany and the UK'. *Transport Policy*. 1 (1), pp. 21-31. And: Reid, S. and Shore, F. (2008) 'Seeing issues clearly - valuing urban realm'. MVA Consultancy for Design for London.
326 See for example: DfT (2006) 'Attitudes to car use'. London: transport statistics. And: Scottish Executive (1999) 'Why people don't drive cars'. Scottish Executive, Central Research Unit.
327 'National travel survey'. Table NTS9902 On: www.gov.uk
328 Clark, B., Chatterjee, K., Melia, S., Knies, G. and Laurie, H. (2014) 'Examining the relationship between life transitions and travel behaviour change: New insights from the UK household longitudinal study'. In: 46th Universities' Transport Studies Group Conference, Newcastle University, 6-8 January.
329 Clark, B., Lyons, G. and Chatterjee, K. (2010) 'The process of household car ownership change: A qualitative analysis of real world accounts'. In: 42nd Universities Transport Study Group Conference, Plymouth, UK, January.
330 DfT (2012) 'National travel survey' Table NTS9902.
331 Delbosc, A. and Currie, G. (2013) 'Causes of youth licensing decline: A synthesis of evidence'. *Transport Reviews*. 33 (3), pp. 271-90.
332 DfT (2012) 'National travel survey'. Table NTS0201
333 A whole edition of the journal Transport Reviews was dedicated to this issue in 2013 Volume 33, Issue 3.
334 Stokes, Gordon (2013) 'What is "peak car"'? On: www.gordonstokes.co.uk

14 London: the politics of bucking the trend

335 Transport for London (2012) 'Travel in London'. Report 5.
336 Note on naming: 'London Transport Executive' was the legal name for the transport agency of the GLC. Its public-facing brand name was 'London Transport', which is how most people, including Dave Wetzel in the following interviews, referred to it. In 1984 when the Thatcher government removed responsibility from the GLC, it was renamed 'London Regional Transport' then 'Transport for London' when the elected mayor assumed responsibility for it in 2000.

337 Though I don't have the references any more, I remember two surveys reported by the Institute of Personnel Management in the mid-1980s. One cited 'rising transport costs' as one reason for the 'flight from the cities'. Another showed London as one of the least favoured places to work among young people.
338 Business rates (or 'non-domestic rates' as they are legally known) are a local tax levied on businesses in the UK. Local authorities used to decide on the rate of the tax in their area, but this power was later removed by the national government. Business rates are now set by national government, pooled and redistributed to local authorities according to a formula.
339 *Urban Public Transport Factbook* on www.publicpurpose.com. Original sources DfT and TfL.
340 Wolmar, Christian (1993) Travelcard 'to vanish with bus deregulation': Leaked report sets scenario for public transport users in London
341 Office for National Statistics (2013) 'Annual mid-year population estimates'. On: www.london.gov.uk
342 On: www.citymayors.com 'The largest cities in the world by land area, population and density'.
343 On: http://legacy.london.gov.uk
344 TfL (2011) 'Travel patterns and trends in London' Table 3. On: data.london.gov.uk
345 Livingstone, K. (2012) *You Can't Say That – Memoirs*. London: Faber and Faber. p. 508.
346 Livingstone, K. (2012) *You Can't Say That – Memoirs*. London: Faber and Faber. p. 396.
347 National Audit Office (2009) 'Report for the Department of Transport: the failure of Metronet'. On: www.nao.org.uk
348 Livingstone, K. (2012) *You Can't Say That – Memoirs*. London: Faber and Faber. p. 473
349 TfL (2011) 'Travel in London: Supplementary report – London travel demand survey'. Tables 7.1 and 7.2
350 Beecham, R. and Wood, J. (2013) 'Exploring gendered cycling behaviours within a large-scale behavioural data-set'. *Transportation Planning and Technology*. pp. 1-15.
351 KPMG (2009) 'Independent strategic review of the provision of bus services in London'. Transport for London. On: www.tfl.gov.uk
352 TfL (2011) 'Travel patterns and trends in London' Table 3. On: data.london.gov.uk
353 TfL (2012) 'London overground impact study: Report to the Board of TfL'. 2 February. On: www.tfl.gov.uk/assets
354 TfL (2013) 'Roads Task Force technical note 1: What is the over-arching transport and travel context to which the Roads Task Force needs to have regard?' On: www.tfl.gov.uk
355 See for example: Bates, J. and Leibling, D. (2012) 'Spaced out: perspectives on parking policy'. RAC Foundation. On: www.racfoundation.org
356 TfL (2011) 'Travel in London supplementary report: London travel demand survey'. Table 7.4.
357 From the 1991, 2001 and 2011 censuses – Table KS17
358 Melia, S., Barton, H. and Parkhurst, G. (2013) 'Potential for carfree development in the UK'. *Urban Design and Planning*. 166 (2), pp. 136-145. On: eprints.uwe.ac.uk/
359 TfL (2008) 'Congestion charging impacts monitoring'. Report: 6.Transport for London.
360 Buckingham, C., Doherty, A. R., Hawkett, D. C. L., Vitouladiti, S. (2010) 'Central London congestion charging: understanding its impacts'. *Proceedings of the ICE - Transport*, 163 (2) 1 June pp. 73–83
361 Givoni, M. (2011) 'Re-assessing the results of the London congestion charging scheme'. *Urban Studies*.
362 TfL (2013) 'Annual report and statement of accounts 2012/13'. On: www.tfl.gov.uk
363 TfL (2013) 'Annual report and statement of accounts 2012/13'. Compared to: KPMG (2009) 'Independent strategic review of the provision of bus services in London'. Transport for London. On: www.tfl.gov.uk
364 Parliamentary Scrutiny Unit (2012) 'Differences in public sector transport spending across England'. On: www.parliament.uk
365 Singapore Government Land Transport Authority (2011) 'Passenger Transport Mode Shares in World Cities'. *Journeys*. 7 November. On: http://ltaacademy.gov.sg
366 See examples on: http://hackney-cyclists.org.uk
367 TfL (2013) 'The vision and direction for London's streets and roads'. Roads Task Force Report. On: www.tfl.gov.uk
368 Carslaw, David (2014) 'Recent findings from comprehensive vehicle emission remote sensing measurements 2014'. London Air Quality Network Annual Conference, 23-24 June. London. On: www.londonair.org.uk

15 Progress in other British cities

369 Although not directly comparable, 'National travel survey' (2012) table NTS9902, shows car ownership falling slightly in the metropolitan areas, and more so in smaller cities since 2005.
370 Edinburgh City Council (2013) 'Transport and travel: Topic summary of car ownership, mode of travel to work and mode of travel to study'. On: www.edinburgh.gov.uk
371 ONS (2013) 'Census consultation: future provision of population statistics'. On www.ons.gov.uk
372 Birmingham, Bristol, Leeds, Liverpool, Manchester, Newcastle, Sheffield and Nottingham.
373 French, Roger (2010) *Pride and Joy*. Brighton: Best Impressions.
374 Census Tables CT0015, KS404 (2011) and L82, L21 (1991). Note that the method of counting home workers changed in 2011, so these figures, which entirely exclude home workers, may differ from others which you will find elsewhere.
375 Trimington, Adam (1996) 'Traffic blow as road hits dead end'. *Brighton Argus*. 29 November.
376 Gould, Christopher (2002) 'Letter to the editor'. *Evening Argus*. 19 July.
377 French, Roger (2010) *Pride and Joy*. Brighton: Best Impressions. p. 44
378 Chiles Andy (2009) 'Has the A259 bus lane worked?' *Brighton Argus*. 11 September.
379 French, Roger (2010) *Pride and Joy*. Brighton: Best Impressions.
380 See the map on: www.brighton-hove.gov.uk
381 *Brighton Argus* (2001) 'No room for residents parking'. 7 March.
382 *Brighton Argus* (2005) 'Business backs park-and-ride'. 13 December.
383 Parkhurst, G. (1995) 'Park and ride: Could it lead to an increase in car traffic?' *Transport Policy*. 2 (1), pp. 15-23.
384 Advertising Standards Authority (2014) 'ASA adjudication on unchain the Brighton motorist'. February 19th. Complaint ref: A13-243394. On: www.asa.org.uk
385 Sage, J., Smith, D. and Hubbard, P. (2012) 'The rapidity of studentification and population change: there goes the (student)hood'. *Popul. Space Place*.
386 Census Tables CT0015, KS404 (2011) and L82, L21 (1991). Note that the method of counting home workers changed in 2011, so these figures, which exclude home workers, may differ from those you find elsewhere.
387 Direct communication, Brighton and Hove council. The patronage for 1992-3 showed 22.5m journeys, for 2010-11 43m.
388 University of Sussex (2009) 'Travel plan 2009–2015'. On: www.sussex.ac.uk And: Colin Buchanan (2009) 'University of Brighton travel plan strategy'. On: www.brighton.ac.uk
389 ONS (2012) 'Annual population survey, commuter flows, local authorities in Great Britain, 2010 and 2011. On: www.ons.gov.uk
390 The comparisons of commuting flows were all obtained from: census.ukdataservice.ac.uk
391 ONS (2014) Table QS702EW - Distance travelled to work. On: www.nomisweb.co.uk
392 Marrs, Colin (2012) 'How we did it: Deal locating greenbelt land'. *Planning Magazine*. 2 November.
393 www.home.co.uk showed average price increases from April 2000 to December 2012 of 108 per cent for Cambridge and 123 per cent for Cambridgeshire.
ONS (2013) House Price Index, November 2013: Table 24 showed increases of 130 per cent for East Anglia and 138 per cent for the UK from 2000 to 2012.
The following article in the FT quotes research by Savills, which likewise shows increases in Cambridge similar to the ONS national average: Bloomfield, Ruth (2013) 'Cambridge closes the gap on Oxford over property prices'. *Financial Times*. 9 August.
394 www.home.co.uk
ONS (2013) House Price Index, November 2013: Table 24 showed increases of 130 per cent for East Anglia and 138 per cent for the UK from 2000 to 2012. The following article in the FT quotes research by Savills, which likewise shows increases in Cambridge similar to the ONS national average: Bloomfield, Ruth (2013) 'Cambridge closes the gap on Oxford over property prices'. *Financial Times*. 9 August.
395 DfT (2013) 'Local area walking and cycling statistics: England 2011/12'. On: www.gov.uk
396 Menzies, B. (2002) 'The Cambridge access strategy'. Association for European Transport Conference. Cambridge, 9-11 September.
397 Cambridgeshire county council (2013) 'Traffic monitoring report 2012'. On: www.cambridgeshire.gov.uk
398 Cambridge News (2013) 'Railings plan to end conflict between cyclists and pedestrians on Cambridge bridge'. 19 March. On: www.cambridge-news.co.uk
399 Cambridge Cycling Campaign (1998) Newsletter 21. National cycle campaigners' conference. December.

On: www.camcycle.org.uk
400 Cambridgeshire County Council (2011) 'Cambridge central area quality bus partnership'. Report to meeting of the Cambridge Environment and Traffic Management Area Joint Committee, 10 October. On: www.cambridgeshire.gov.uk
401 Havergal, Chris (2013) 'Guided busway legal battle with BAM Nuttall settled by Cambridgeshire County Council'. *Cambridge News*. 30 August.
402 Atkins (2012) 'Cambridgeshire guided busway post-opening user research for Cambridgeshire county council'. On: www.brtuk.org
403 See for example: www.castiron.org.uk
404 See for example: Currie, G. (2005) 'The demand performance of bus rapid transit'. *Journal of Public Transportation*. 8 (1), pp. 41-55.
405 Cambridgeshire County Council (2012). 'Transport strategy for Cambridge'. On: www.cambridgeshire.gov.uk
406 Havergal, Chris (2014) 'Is shutting cars out of more of central Cambridge the solution to the city's traffic problems?' *Cambridge News*. 9 February.
407 Melia, S. (2011) 'Students car use and its effect on environmental attitudes'. Working paper. University of the West of England, Bristol.
On: eprints.uwe.ac.uk/14424/
408 Melia, S. (2014) 'Do randomised control trials offer a solution to "low quality" transport research?' Working paper. University of the West of England, Bristol.
On: eprints.uwe.ac.uk/16117/
409 www.demographia.com 'International: Selected cities with declining population ranked by annual loss rate'
410 Merseytravel (2006) 'Second local transport plan for Merseyside 2006 – 2011'.
On: www.knowsley.gov.uk
411 DfT (2014) Table CGN0201a 'Average vehicle speeds (flow-weighted) during the weekday morning peak on locally managed 'A' roads by local authority in England, annually from 2006/07'. On: www.gov.uk
412 Population, driving to work and car ownership from: Census Tables CS1, CS121, CT00015, KS404EW. Household income from: www.gov.uk 'Gross disposable household income by sub-region' Table 3.1 (NUTS Level 1), adjusted for RPI changes at January each year. As income figures were shown at County level, Cambridgeshire is used instead of Cambridge. Car traffic volumes from DfT Table TRA8902 except Cambridge, which is taken from: Cambridgeshire County Council (2013) 'Traffic monitoring report 2012'. Table 3.4 'Traffic growth on the Cambridge radial cordon 2002 – 2012'.

16 What sort of cities do we want?

413 DfT (2014) Table CGN0201a 'Average vehicle speeds (flow-weighted) during the weekday morning peak on locally managed 'A' roads by local authority in England, annually from 2006/07'. On: www.gov.uk
414 See for example: Cassini, Martin (2012) 'Viewpoint: Is it time to get rid of traffic lights?' *BBC News Magazine*. 16 May. On: www.bbc.co.uk
415 There is another myth around this question, which runs as follows: modern agricultural practices are environmentally damaging; there is more biodiversity in the average suburban garden; therefore we can improve the environment by building on farmland. There is a lot of truth in the first of these statements, some truth in the second, none whatsoever in the third.
416 Bates, J. and Leibling, D. (2012) 'Spaced out: perspectives on parking policy'. RAC Foundation.
417 CLG (2013) 'Eric Pickles launches package of support for local shops'. Press release. 6 December. On: www.gov.uk
418 Liberal Democrats (2013) 'Green growth and green jobs. Transition to a zero carbon Britain'. Policy paper 109.
On: www.davidgoodall.org.uk
419 Grayling, T., Sansom, N., & Foley, J. (2004) 'In the fast lane: fair and effective road user charging in Britain'. London: Institute of Public Policy Research.
420 See for example: Downs, A. (2004) 'Why traffic congestion is here to stay.... and will get worse'. *Access*. 25 (Fall), pp. 19-25. On: www.uctc.net
421 CBI (2012) 'Bold thinking: road report'. Confederation of British Industry. On: www.cbi.org.uk
422 HM Treasury (2013) 'Investing in Britain's future'. London. On: www.gov.uk
423 Bhatta, S. D. and Drennan, M. P. (2003) 'The economic benefits of public investment in transportation: A review of recent literature'. *Journal of Planning Education and Research*. 22(3), pp. 288-96.
424 Dumbaugh, Eric (2012) 'Rethinking the economics of traffic congestion'. *Citylab*. 1

June. On: www.citylab.com
425 ONS (2012) 'Regional GVA NUTS3 official statistics showing annual estimates of NUTS3 regional gross value added'. December. Table 3.3. And: DfT (2013) 'Road congestion and reliability statistics'. Table Table CGN0203a. Both available on: www.gov.uk
426 ONS (2013) 'National population projections 2012-based extra variants report'. On: www.ons.gov.uk
427 Intensification in the centre of Liverpool was offset by population reductions in other areas.
428 Tallon, A. and Bromley, R. (2004) 'Exploring the attractions of city centre living: evidence and policy implications in British cities'. *Geoforum*. 35 (6) pp. 771–87 found this in Bristol. I found it in Inner London: Melia, S. (2009) 'Potential for carfree development in the UK'. Unpublished PhD, University of the West of England. On: www.stevemelia.co.uk.
430 See www.populationmatters.org for more about population issues.
431 Northstowe Joint Development Control Committee (2012) 'Northstowe development framework document'. On: www.scambs.gov.uk
432 Hallam Land Management, Persimmon, Redrow and Taylor Wimpey (2010) 'East Devon new community phasing strategy'. On: www.eastdevon.gov.uk.
433 ONS (2013) 'National population projections 2012-based extra variants report'. On: www.ons.gov.uk
434 See for example: *Bristol Post* (2014) 'Protesters angry as controversial plans for a MetroBus route for north Bristol narrowly approved'. 28 August. On: www.bristolpost.co.uk
435 Melia, S., Barton, H. and Parkhurst, G. (2013) 'Potential for carfree development in the UK'. *Urban Design and Planning*. 166 (2), pp. 136-45. On: eprints.uwe.ac.uk/
436 Douglas, N.J. and Jones, M. (2013) 'Estimating transfer penalties and standardised income values of time by stated preference survey'. In: Anon. (2013). '36th Annual Conference of the Australian Transport Research Forum'. Brisbane, 2-4 October. On: www.atrf.info
437 'National travel survey'. Table NTS9902 On: www.gov.uk
438 Harmer, C. and Cairns, S. (2011) 'Carplus annual survey of car clubs 2010-11'. On: www.carplus.org.uk
439 I will be doing more research on car clubs in carfree and low-car developments with Carplus, the national umbrella organisation for car clubs, in the near future.
440 See: Hickman, Kevin (2014) 'Do we need cycling behaviour change for planners, advocates and policy-makers?' Presentation to Cycling and Society Symposium. Newcastle. September.
441 MacKay, D.J.C. (2009) *Sustainable Energy without the Hot Air*. Cambridge: UIT. Also available free on: www.withouthotair.com
442 This was explained to me by a contact in the research and development division of a major car manufacturer.
443 Rice, Craig (2014) 'Barriers to electric car take-up'. *The Engineer*. 17 March. On: www.theengineer.co.uk
444 See: Scheuer, Jeffrey (2013) *The sound bite society: How television helps the right and hurts the left*. New York: Routledge.
445 DfT (2008) 'Carbon pathways analysis: Informing development of a carbon reduction strategy for the transport sector'. Available on the National Archives website.
446 Begg, D. (2014) 'A 2020 vision for London: What are the implications of driverless transport?' *Clear Channel*. On: www.transporttimes.co.uk

17 What can I do?

447 Lessing, Doris (1978) *The temptation of Jack Orkney: Collected stories volume 2*. London: Jonathan Cape. p. 176.
448 Photographs reproduced courtesy of Chris Todd and the South Downs Campaign.
449 Tyndall Centre (2013) 'Climate transitions: connecting people, planet and place'. Annual PhD Conference. 3-5 April. Cardiff University.
450 Melia, S (2010) 'My carfree journey'. *AtoB Magazine*. 78, June. On: www.stevemelia.co.uk
451 Melia, S. (2014) 'Sustainable travel and team dynamics amongst mobile health professionals'. *International Journal of Sustainable Transportation*. ISSN 1556-8334
452 The original Grand European Tour was largely confined to a wealthy minority; a similar idea could apply to a much wider group of people today.

Index

Page numbers in *italic* refer to figures

A
accessibility planning 222
air pollution 25, 30, 42, 227
airport expansion 33–4, 110
Amsterdam *62*, *63*, 67, 148, 158, 217
anti-collision technology 119
Ashford 81, 84, 85–6
Athens 227
autonomous vehicles 119, 214, 229–30
aviation
    air travel restraint policy 34
    airport expansion 33–4, 111
    business travel 10, 238
    greenhouse gas emissions 13, *14*
    leisure flights 29, 33, 238
    'tourism gap' 29, 33–4, *34*

B
Baker, Norman 22
Barclays Cycle Superhighways 176
Basel 61, *62*, 148
Bassam, Steve 185, 232
Bath 217
Baugruppen 157
Bayliss, David 165
Berkeley Group 56–7, 58
Big Brother Watch 27
Birmingham *99*, 180
Bogotá 71
Boles, Nick 93
Bonn 63
Bordeaux 148
Bracknell 151
Bradley Stoke 94–5, 96, *96*, 105, 107–8
Brighton 85, *86*, 184–95, 221
    bus use 186–8, *186*, *189*, 193, 194
    car ownership *210*
    car-free development 189, *189*
    cycling 190, *191*, 194
    demographics 189, 192–3, *210*
    Green Party policies 190–2
    housing development 189, 195
    modal shift 192–4
    park and ride 190–1
    parking schemes 188
    rail 194
    travel patterns *185*, 192–4, *193*
    walking 194–5
Bristol
    car-free development 161
    cycling 63, *67*, 72, 76, 78, *78*
    housing development 99, *99*
    pedestrianization *88*, 208
    road network 116–17, *117*
    trams 141
    UWE transport 206–8
Bristol to Bath cycle path 72, *73*, 74
British Social Attitudes Survey 59
Brown, Gordon 104, 173, 174
brownfield development 55, 98–9, 100, 101
Brundtland Commission 120–1
Buchanan report 112–16, 117–18, 150, 164, 182
bus gates 64
bus travel
    bus rapid transport (BRT) 40, 205, 211, 221
    bus-to-bus changes, resistance to 41, 222
    competitive practices 45–6
    deregulation 23–4, 38, 43, 45–6, 193
    diesel buses 42
    European countries 43
    fares 23–4, *24*, 38, 44
    free 41–2, 43
    guided busways 90, *202*, 204–5, *205*
    impact on car use 38, 41, 42, 106
    low emission buses 42
    national bus use *44*
    orbital bus services 41
    peak-time commuting 42, 43
    in permeable street networks 90
    radial routes 41
    real-time information 188
    reregulation 223
    speeds 41
'bus wars' 45
business attitudes 134–5, 143–4, 147, 182, 191–2
business travel *see* work, travel to
bypasses 113

C
calculation fallacy 10–11, 37
Cambourne *105*, 196
Cambridge 60, 76, 184, 192, 195–206, 219
    bus use 204–5, 206
    Cambridge Core Traffic Scheme 197–200, *197*, 202, 204, 221, 235
    car ownership *210*
    cycling 196, 200–4, *201*, *202*, *203*
    demographics *210*
    filtered permeability 202, *203*
    Holford report (planning strategy) 195–6
    political consensus 206
    public transport 204–6
    rail 204
    travel patterns 185, *200*
Campaign to Protect Rural England (CPRE) 17–18, 232
car clubs 52, 155, 159, 222, 224, 227
car ownership
    Buchanan report forecasts 112–13, *112*
    demographics 162, 222
    income and 53, 57
    'normality' of 26, 28, 238

car ownership *cont.*
  parking restraints, impact of 49, 54, 55, 57–9, 110, 218, 219
  predictor of car use 49, 50–3, *50*, 229
  public transport, impact of 58
  reasons for not owning 53–4, 57, 180
  rural areas 224
car purchase tax 21
car sharing 52
car use
  adults without cars 52–3, 52, 229
  cost trends 21, *21*, 24, *24*
  income and 53
  measuring 51
  'peak car' 28, 211
  personal freedom issues 25
  public transport improvements, impact of 38, 39, 110
  social costs 25
  time savings 35
  see also shared space
car-free choosers 27, 28, 53–4, 161–2, 237–8
car-free developments 149–62, 219–20
  parking overspill 157–8, 160
  social benefits 159
  see also under Brighton; Cologne; Freiburg; London; Louvain-la-Neuve
carbon emissions 42, 119, 229
carbon trading systems 37
Cardiff 184
cars
  carbon emissions *14*, 16, 19, 24–5
  environmental-related problems 16
  see also autonomous vehicles; electric cars
casualties
  cycling and 27–8, 67, 69, *70*, 75–6, 78
  legal liability for 75–6
  pedestrian 79
  speed and 26, 27, 36
catchment populations 96

Churchill, Winston 24
climate change 13, 14, 110, 229, 236, 237
  cost-benefit analysis (CBA) and 37
Climate Change Act 2008 14
Climate Change Committee 14, 15, 226, 229
cognitive dissonance 237
Collomb, Gérard 144
Cologne 148
  Stellwerk 160 (car-free development) 158–61, *158*, *159*, 160
combined local authorities 181
commercial development 55, 56
Commission for Integrated Transport (CfIT) 50–1
Competition Commission 45
Confederation of British Industry (CBI) 30, 215
congestion 42, 211–14, 229
  economic costs 215
  options for dealing with 113–22, 212–14, 229–30
  rich cities 215
  tolerance of 118, 214
congestion charging 77, 122, 171, 174–5, 180, 181, 213, 214
connectivity 146, 222
Connex 46
Copenhagen 61, 69, 148
cost-benefit analysis (CBA) 29, 30, 35–7, 146, 228
courtesy crossings 86
Coventry 63
Cranbrook 107, 220
Crossrail 174, 178, 225
cycle hire schemes 71, 143, *143*
cycle parking 68, *69*, 130, *131*, 135, 139, *140*, 159
cycling 23, 60–78, 224–6
  cycling culture argument 61, 70
  European rates of 60, 61, *61*, 62, *62*, 63, 68–9, 182
  flat terrain and 63, 70
  government spending on 73
  health benefits 30
  hostility to 71, 225, 226
  main roads 76
  male dominance (UK) 67, 73,
75, 176, 226
  pedestrianization and 67, *68*, 87
  in permeable street networks 90
  protective clothing 67
  push factors 70
  reasons for not cycling 65, 67, 75
  red lights, jumping of 68
  and road casualties 27–8, 67, 69, *70*, 75–6, 78
  shared-space schemes 88
  time cost 75, 76
  traffic conditions and 63, 75
  UK rates of 61, 71–2, 76
  weather conditions and 63
Cycling Embassy of Denmark 77
Cycling Embassy of Great Britain 77, 203–4
cycling networks 37, 60, 61, 64, 71, 72
  ad hoc design compromises 75
  cycling contraflows 67, 176, *176*
  cyclist-only traffic light phasing 68, 139, *140*
  'Dutch-style' infrastructure 77–8, 225, 226
  European cities 64–5
  filtered permeability 64, *65*, 176, *176*
  hierarchy of provisions 74–5, *74*, 77, 225
  hostility to segregation 72, 73
  hybrid paths 65, *66*, 75, 77, *191*
  priority at junctions 68
  safety auditors 65
  segregated routes 71, 72, 73, 75, 76, 77, 82
  selective road closures 67
  separation of cyclists and pedestrians 65, 78, 225, 226
  substandard provision 76, *76*
cycling priority roundabouts 65, *66*
Cyclists Touring Club (CTC) 71–2, 77

D

Darlington Transport Company (DTC) 45
Daseking, Wulf 124
Davey, Ian 190, 191, 192
De Mello, Lianne 172, 191–2
Delleske, Andreas 157
demarcations, removal of 81, 85
Denmark
 cycling 61, 65, *66*, 69, 75, 78
 see also Groningen
Department for Transport (DfT) 11
 cost-benefit analysis (CBA) 35–7
 and eco-towns, transport in 106
 Economic case for HS2 47–8
 Manual for Streets 75, 89, 91
 National Cycling Strategy 72
 and shared space 80, 84, 89
Department of the Environment, Transport and the Regions 98
deregulation 23–4, 38, 43, 45–6, 84, 168, 181, 193
Dickens Heath *105*
diesel buses 42
disabled people 84, 86, 114, 139, 226
domestic tourism 34
Downs, Anthony 214
Drachten 82, *82, 83*
driverless cars *see* autonomous vehicles
Dutch Cycling Embassy 76–7

E

Earl, Dave 197, 198, 200–1, 204, 206, 235
eco-towns programme 104–5, 220, 221
 opposition to 106
 transport facilities 106
Edinburgh 158, 161, 184
electric cars 14–15, 110, 119, 224, 226–7, 229
 charging systems 15–16, 227
environmental campaigns
 local campaigning 231–5
 personal travel behaviour 236–8
 transport planners 235–6
environmental capacity 114
environmental sustainability, attitudes to 110
European best practice 120–2
 see also Freiburg; Groningen; Lyon
European travel patterns 120, *121*
Exeter 161, *162*
Exhibition Road, London 80, *80*, 219

F

false consensus effect 56
falsely positive responses to surveys 207
'family houses' 103, 218
Ferguson, George 208
ferry services 34
filtered permeability vii, 64, *65*, 90, 106, 113, 137, 175, *176*, *176*, 182, 197, *199*, 202, *203*
Filton 95–6, *96*
First Group 47
flats 10, 98–9, *100*, 218
 proportion of housing stock 92, 100, 101, 102, *102*, 103
 social housing 103
'floating bus stops' 190, *191*
flyovers 114
footbridges 64
France
 local government 145
 motorway networks 120
 national planning policy 144–5, 227
 tram systems 221
 transport revenue stream 228
 see also Lyon
free market ideology 84
Freiburg 61, *62*, 63, 64, 69, 122, 123–34, 219, 221
 car ownership 132, *133*, 157
 cycling network 129–30, *130, 131*
 demographic profile 148
 parking 132, 155, 157
 pedestrianization 124, *124, 133*
 public transport, extension of 125–8, *125*
 road network *128*
 stellplatzfrei residential streets 155–6, *156*
 traffic restraint 128–9, 147
 trams 126, *126*, 127–8, *127*, 132, 141
 transport plan 123, 124–30
 travel patterns 132, *133*
 urban planning 130–2
 Vauban (car-free development) 155–8, *156*
 walking, fall in 147
French, Roger 187, 188, 193
fuel tax 10, 19, 20–1, *21*, 213, 214
fuel tax protests 20, 213

G

garden cities 106–7, 220
 funding 106–7
 see also eco-towns programme
Germany
 car ownership and use 10, 49, 51, *51*, 52
 cycling 61, 63, 65, 69
 motorway networks 120
 transport decision-making 227
 see also Freiburg
Glasgow 184
GNER 47
good practice
 British *see* Brighton; Cambridge; London
 European *see* Freiburg; Groningen; Lyon
governance issues 227–8
GPS route finding 118
greenfield development 17, 18, 55, 97, 100, 117, 212, 216–17, 220, 221, 234
 land prices 103–4
greenhouse gas emissions 13–14, *14*, 42, 119, 229
grid roads 89, *89*, 117, *118*
Groningen 61, *67*, 122, 134–41, 147, 219
 car ownership 134, 136, 140
 cycle network 64, 137–9, *138*
 demographic profile 148
Groningen *cont.*
 parking 136

traffic restraint 134-6, 136, 147
travel patterns 139, *140*
Gross Domestic Product (GDP) 30
guardrails 81
guided busways 90, *202*, 204-5, *205*

H

Hackney 175, *176*, 182, 219, 225
Hammersmith Flyover 116
Hammond, Philip 22
Harlow 151
Hasselt 41-2
health and safety culture 65
heavy goods vehicles (HGVs) 121
Heffer, Simon 27
Heidelberg 63
Hendy, Peter 163, 166, 169, 171-2, 177, 182
high-speed rail (HS2) 22, 23, 37, 38, 47, 223
  budgeted cost 47
  customer projections 40, *40*
  low-carbon argument 47-8
home zones 156
Hong Kong 23
household size and composition 10, 17, 92, 101, *102*, 106
housing crashes 94
housing densities 94, 96-7, 98, 99, 100, *101*, 106, 151
housing development
  brownfield development 55, 98-9, 101
  'car-free' 57, 58
  density guidance 98, 99
  facilities, provision of 95
  'fill in and spread out' 117, 216
  greenfield development 17, 18, 55, 97, 100, 101, 103-4, 117, 212, 216-17, 220, 221, 234
  'knock down and spread out' 116, 118, 164, 212, 216
  localised employment opportunities 107-9
  parking requirements and guidelines 55, 56, 98, 99
  see also eco-towns programme; garden cities; New Towns programme; small new settlements; urban intensification
housing prices 218
hybrid paths 65, *66*, 75, 77, *191*
hydrogen 16

I

immigration 17
in-group favouritism 19, 26, 27
integrated transport networks 36-7, 71, 146, 147, 214, 216, 227
  joined-up decision-making 227-9
inter-urban travel 117
International Air Transport Association (IATA) 33
Islington 57
Ivybridge 65, 105, *105*, 107, 108-9, *108*, 234

J

jet packs 118
Johnson, Boris 77, 174, 176-7, 178

K

Kent County Council 56, 177
Kent Fastlink 40, *40*
kerbs 81
Kiley, Bob 173, 175
Klaauw, Cor van de 137, 139
Kleinmann, Hans-Georg 158, 159

L

land prices 103-4
land use planning
  eco-towns programme 104-5, 220, 221
  New Towns programme 94, 103, 104, 117, 150-1, 154, 221
  post-Second World War 94
  relationship to transport 92-109
  see also housing development
Lessing, Doris 231
Leuven 149
light rail networks 37, 141
Liverpool 208-9, *209*, *210*
Livingstone, Ken 163, 164, 165, 166-8, 169-70, 171, 173, 174-5, 176-7, 178, 181, 182, 183
Local Enterprise Partnerships 228
Local Sustainable Transport Fund 22
Local Transport Today 13, 235
localism 100
London 163-83
  average traffic speeds 213
  bus use 39, 42, 44, 46, 163, 168, 171-3, 181
  car clubs 224
  car ownership 58, 179-80, *180*, 210
  car use 62
  car-free developments 161, *162*, 179, 220
  car-free households 58
  commuting into 108, 194, 204
  congestion charging 77, 122, 171, 174-5, 180, 181, 213
  cycling 62, *62*, 77, 163, 175-6, *176*, 182, 224-5
  demographics 178, 179, *210*
  Fares Fair policy 165-6, 171
  GLC, abolition of 167, 168
  Mayor's Transport Strategy 170-1
  Oyster card 44, 172, 173
  parking policies 56-7, 58, 179
  population 58, 169, *170*
  public transport network 62, 163-83
  public transport subsidy 165, 181, 182
  rail 177, 178
  Roads Task Force 182, 220
  shared space 80, *80*
  total traffic volumes 163, 180
  trams 141
  transport issues (1980s) 163-8
  transport policy under Boris Johnson 176-7, 178
  transport policy under Ken Livingstone 163-76, 178
  travel patterns *164, 167*,

178–9
  Travelcards 165, 167, 168
  underground travel 39, 62, 171, 173–4, 177
  walking 163, 182
London Cycling Campaign 77, 175–6
London First 169
London Overground 177, 178
London Plan 169
London Regional Transport (LRT) 167, 168, 169
Louvain-la-Neuve 147, 148, 149–55, *150*, 219
  car ownership 152
  car-free development 149–55
  cycle network 150, 154–5
  housing density 151
  parking 152
  rail travel 151–2, 154
  shared space 150
  travel patterns 152–3, *153*, 154
  walking 152, 153, 154
Lyon 122, 141–6, 147, 219
  capital investment programme 147
  car ownership 145
  cycle hire 143, *143*
  cycle network 143, *144*
  demographic profile 148
  parking 143–4
  public transport budget 141–2, *142*
  traffic restraint 147
  trams 141, *142*, *144*
  transport tax 142
  travel patterns 145–6, *146*
  trolleybuses 141, *142*
  walking, growth in 147
Lyons Review 107

M
Maastricht 148
Major, John 72, 168
Malmö 61, *65*, 148
Manchester
  bus travel 40, 41
  Metrolink 40, *40*, *41*, 126–7
  trams 10, 40, 126–7, 181–2
Marples, Ernest 112
Menzies, Bob 199
Merseyrail 209, *209*

Milan 175
Milton Keynes 117, *118*, 151, *152*, 221
Mitchell, Gill 185, 186, 187, 189, 190, 195
mobility scooters 139, 226
modal shares 62, *74*, *95*, 121, *133*, 134, *140, 146, 154, 164*, 184, *185*
modal shift 20, 48, 83, 87, 147, 178, 180, 192–4, 205, 207, 221
mode neutrality 20
Monderman, Hans 82, 83
Monopolies and Mergers Commission 45
Moody, Simon 84
Morton, Alex 93
motorway tolls 21
multi-level roads 115–16
multi-storey parking 115, 159
multiplier effects of public spending 31
Münster 148
MVA Consultancy 83, 85
myths of urban transport 9–10, 19, 29, 38, 49, 60, 79, 110

N
Nailsea 151
Napier, Christopher 233–4
National Bus Company (NBC) 186, 193
National Cycle Network 65, 72
National Cycling Strategy 72, 73–4, *74*
National Express 47
national parks 189, 191, 232–3
national pattern of transport movements 39, *39*
national transport budget 73
Nelson 87–8
neoliberal ideologies 84, 93, 174
Netherlands
  cycling 60, *61*, 63, 64, *65*, 68, 69–70, *70*, 75, 78, 82
  motorway networks 120
  shared-space schemes 85
  transport decision-making 69, 227
  see also Groningen
Network Rail 47

New Towns programme 94, 103, 104, 117, 150–1, 154, 221
New Urbanist movement 75, 91
New York City 57
nimbyism 17, 97, 234
Norris, Stephen 177
North American street layouts 89, *89*, 90
Northampton 63
Northstowe 220
Nottingham 193, 228
Nuremberg 156

O
Odense *62*, *68*, 69, 148
oil crisis (1973–4) 69
out-group behaviour 27
Oxford 191
  bus travel 42, *43*, 193
  cycling 196
Oyster card 44, 172–3

P
Paris 170, 182, *183*, 217, 227, 239, *239*
park and ride schemes 42, 190–1, 197
parking
  allocated parking 56
  charges 129, 143, 147, 207, 213
  commercial car parks 219
  cycle parking 68, *69*, 130, *131*, 135, 139, *140*, 159
  driverless 119
  multi-storey 115, 159
  new housing developments 55–6
  Radburn layout 151
  underground parking 143–4
  workplace parking levy 228
parking permits 57, 152, 188
parking restraint
  impact on car ownership 49, 54, 55, 57–9, 110, 218, 219
  local authority discretion 56
  new developments 55–6
  opposition to 221
  without parking control 55, 56, 57, 100
Parsons, Trevor 175
peak-time travel 42, 114, 172,

223
pedestrian crossings 219
pedestrian walkways 114–15, *115*, 116
pedestrianization 32, 79, 80, 81, 87–8, 161, 213
   Buchanan report and 113
   economic impact 87, 161
   night time use 87
   residential developments within 87
   social value 88, 219
   walking and cycling, encouragement to 87
Penn Center, Philadelphia 115, *116*
permeable street networks 79, 89–91
personal freedom issues 25
personal rapid transport systems 119, *120*, 230
personal travel behaviour 236–8
planning permission 100
play streets 128, 156, *157*
Plowden, Ben 178, 179, 182
Plymouth 107, 108, *108*, 109, 234
Policy Exchange 93, 94
policy response bias 134–5
pool cars 238
Poole 54, *54*, 57
population growth 17, 220
Portsmouth 178
Prescott, John 20, 98, 173, 233
Prince's Foundation 89, 91
productivity measurement 31
public realm improvements 143–4, 208, 225
public spending, multiplier effects 31
public transport
   and behaviour change 38, 40, 41–2
   extra journeys, creation of 38, 42, 132, 191
   impact on car ownership 58
   impact on car use 38, 39, 110
   journey times 222
   limitations of 38–48
   promotion costs 147
   rural areas 224
   subsidies 114, 165, 181, 182, 223
public-private partnerships (PPPs) 173, 174

Q
Quality Partnerships 204

R
RAC Foundation 26, 57
Radburn layout 151
rail
   cost trends 24, *24*
   eco-towns and 106
   expansion 46, 47, 221, 222, 223
   fares 23, 223
   HS2 *see* high-speed rail
   privatisation 23, 46–7
   public subsidy 23, *23*, 47
   renationalisation 46–7
   track charges 47
'raised table' junctions 137, *139*
Rennes 143
'right to buy' legislation 103
ring roads 116
rising bollards 197, *199*
risk compensation theory 84
road building 22, 211
   economic impacts 29, 30–3
   funding 19, 24
   and increased traffic generation 29, 33, 110
   public subsidy 23, *23*
road capacity
   increasing 33, 118, 212
   reducing 67, 84, 213
road closures 32–3, 67, 197, 198, 213, 228
road pricing 114, 118, 119, 211, 213–14
   revenue-neutral 213–14
   see also congestion charging
road tax 19, 24
   see also vehicle excise duty
road widths 67
Roads Task Force 182, 220
Runcorn 151, 154
rural transport 94, *104*, 223–4

S
Salomon, Dieter 124

Scotland
   bus service regulation 223
   housing planning policies 99, 103
season tickets 125, 126
Seoul 32
settlement size, and traffic generation *104*, 105, 218
Seville 60, 70–1
shared space 75, 79–91, *80*, 212, 219
   as alternative to pedestrianization 87
   casualty reductions and 84, 85
   concept originator 82
   courtesy crossings 86
   cyclists and 88
   definitions of 80, 81
   demarcations, removal of 81, 85
   hostile environment 79, 85–6
   impact on car use 84
   impact on traffic speed 85
   permeable street networks 79, 89–91
   reasons for implementation 83
   resistance to 88
   risk compensation theory 84
short-termism 228
showers, workplace 67
Sinden, Neil 17–18, 232, 234
Skelmersdale 154
small new settlements
   and traffic generation 93, 105, 106
   unsustainable development 107, 220
smartcards 44, 172–3
Smith, Jeremy 196, 204, 206
social capital 83
social housing 103, 160, 218
social value 30, 88, 159
SOLUTIONS project 94–6
South Woodham Ferrers 105, *105*, 106
Southampton 129
Spain *61*, 70–1
speed cameras 19, 22, 25–7, 36
   revenue-raising argument 25,

26, 27
Stagecoach 45, 187
Stevenage 151
Stockholm 175
Stopps, Vincent 175
Strategic Rail Authority (SRA) 46–7
subsidies
  public transport 114, 165, 181, 182, 223
  rail travel 23, 23, 47
suburbs
  and traffic generation 93
  travel patterns 94-6, 95, 96, *96*
sustainable development: definitions of 120-1
Sustainable Development Commission (SDC) 25
sustainable transport 121
  opposition to 25
  reduced government support for 22–3
  see also good practice
Sustrans 72, 202
Switzerland 61

T
Thatcher government transport policies 43, 72, 164, 167
time savings 31, 35, 36, 37
Todd, Chris 189, 190, 191, 233
Tokyo 182
Town and Country Planning Association (TCPA) 106
traffic lights 81, 84
  cyclist-only phasing 68, 139, *140*
  switching off 212
traffic management in towns 112–22, 212
  Buchanan report 112–16, 117–18, 150, 164, 182
  opposition to 147
  traffic restraint 16, 42, 114
  see also congestion
traffic speeds
  20 mph zones 26, 27, 191–2
  in shared-space schemes 79, 85

trams 10, 205–6, 221
  Bristol 141
  Freiburg 126, *126*, 127–8, 132, 141
  London 141
  Lyon 141, *142, 144*
  Manchester 10, 40, *40*, 126-7, 181–2
Transport for London (TfL) 62, 169, 171, 172, 173, 174, 177, 178, 179, 181, 220, 224-5
transport planners 235–6
transport tax 142, 228
transport–economy relationship 29–37, 215
Travelcards 165, 167, 168
trip chaining 33
trolleybuses 141, *142*

U
UKIP 25, 71, 206
Ultra personal rapid transit system 119, *120*, 230
underpasses 64, 114, 115
university cities 148, 149–50, 192, 194, 196, 206
  see also individual index entries
University of the West of England (UWE) 58, 88, 206–8
'urban extensions' 105, 220
urban intensification 96–7, 99, 100, 217–20, 229, 234
  car-free development, opportunities for 219–20
  paradox of 96–7, 100, 218
  supportive public transport 222
  visible improvements, importance of 219

V
value judgments 11, 29, 36
vehicle excise duty 19, 24–5, 214
Venice 155
Vienna 158, 159
voluntary behaviour change 114, 119, 212-13

W
walking 62, 72
  health benefits 30
  pedestrianization and 87
  in permeable street networks 90
'war on the motorist' 19, 20, 22, 25, 28, 56
Ward, Steve 206-7
Welsh housing planning policies 103
Wetzel, Dave 163–6, *166*, 167, 171, 172, 173, 174–5, 177
White, Bob 56
Wixams 105, *105*
Woitrin, Michel 149, 150, 151, 155, 219
Wolmar, Christian 168, 182–3
work, travel to 92, 222
  bus 40, 194
  car 10
  cycling 194
  flying 10, 238
  local employment opportunities, impact of 107–8
  new settlements and 105–6, *105*
  population density, relation to 93–4, *93*
  rail 35, 105–6, 194, 204
  tram 40
  two-way commuting *108*, 109
  work–home transition time 109

Y
York 191, 196

Also in the 'without the hot air' series

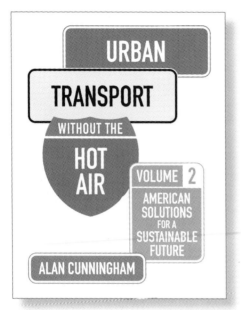

Volume 2 of *Urban Transport - without the hot air looks* at urban transport in the American context. It can be read independently from volume 1, or as a companion to it.

Drawing on his experiences as a planner and the evidence from successful developments, Alan Cunningham proposes a nationwide geographic plan that will allow walking, biking and transit to flourish alongside traffic.

Perfect for people with an interest in the future of transport, this book is also invaluable for anyone involved in the planning and development of urban areas.

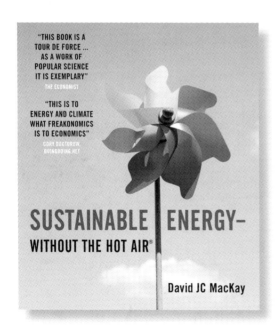

For our full range of titles, see www.uit.co.uk

 @uitbooks

 /UITCambridge